AF251279

Smart Innovation, Systems and Technologies

Volume 219

Series Editors

Robert J. Howlett, Bournemouth University and KES International, Shoreham-by-sea, UK

Lakhmi C. Jain, KES International, Shoreham-by-Sea, UK

Srikanta Patnaik · Tao-Sheng Wang · Tao Shen ·
Sushanta Kumar Panigrahi

Editors

Blockchain Technology and Innovations in Business Processes

 Springer

Editors
Srikanta Patnaik
SOA University
Bhubaneswar, India

Tao Shen
Kunming University of Science
and Technology
Kunming, China

Tao-Sheng Wang
Department of Applied Economics
Hunan International Economics University
Hunan, China

Sushanta Kumar Panigrahi
Interscience Institute of Management
and Technology
Bhubaneswar, India

ISSN 2190-3018   ISSN 2190-3026 (electronic)
Smart Innovation, Systems and Technologies
ISBN 978-981-33-6469-1   ISBN 978-981-33-6470-7 (eBook)
https://doi.org/10.1007/978-981-33-6470-7

This Springer imprint is published by the registered company Springer Nature Singapore Pte Ltd.
The registered company address is: 152 Beach Road, #21-01/04 Gateway East, Singapore 189721, Singapore

Preface

Blockchain is one of the most important technical inventions in the recent years. It is a transparent money exchange system that has transformed the way a business is conducted. It not only reduces risk but also puts an end to fraud along with bringing transparency. Companies and tech giants have started investing significantly in the blockchain market, and it is expected to be net worth of more than 5 trillion dollars in the next 5 years. This distributed ledger technology has increased its prominence in various sectors such as financial services, health care and entertainment. This emerging technology has marked its presence in the government sectors too such as in voting, taxes, identity management and record management. The benefits such as traceability, transparency, reduced costs and increased security make its applications a popular and successful one. This book covers the theory of blockchain technology, its security issues and its application in different industries.

The authors in this volume are from diverse background and from different universities across the globe. The authors of the chapters have related the blockchain technology with energy, supply chain management, data protection and security issues.

I profusely thank all the authors for their valuable contribution without which this volume could not have been possible.

On behalf of the editors, let me thank the Publishing Editor and the team of Springer Nature Singapore for their approval of the book project and constant support and cooperation during the publication.

Last but not least, I would like to extend my heartfelt thanks to my co-editors—Prof. (Dr.) Tao Sheng Wang, Prof. (Dr.) Tao Shen and Dr. Sushanta Kumar Panigrahi—for their constant support in bringing out this volume in time.

I am sure the readers shall derive utmost benefit from this volume.

Bhubaneswar, India   Srikanta Patnaik

Editorial

The evolution of blockchain technology has been marked with bitcoins. Bitcoin is a new kind of money which is being used for innovative payments. However, bitcoin and blockchain are being interchangeably used since bitcoin is a cryptocurrency and blockchain is the underlying technology that has distributed database to empower bitcoin. Again, blockchain technology is considered to be beyond digital money and payments. This technology can change the way that ownership, privacy, uncertainty and collaboration are conceived of in the digital world. Moreover, in several disrupting sectors and diverse practices like financial markets, content distribution, supply chain management, dispersal of humanitarian aid and even for voting in a general election, blockchain technology has huge applicability.

Besides, it provides varied information to the public in general—which has many potential applications that require smart transactions and contracts. As we know, blockchain is said to be immutable since breaking its encryption and decentralized architecture is very complex. Thus, these special features of blockchain technology safeguard the data stored on the shared ledger. Also, with the emergence of blockchain technology, there is a paradigm shift in the way of doing core processes, even for the industries that have already seen significant transformation due to wide adoption of digital technologies. Now, this makes blockchain technology suitable for limitless range of applications across almost all industries. Further, the ledger technology can be applied to most of the sectors, may it be to track fraud in finance or to securely share patient medical records between healthcare professionals. It also acts as a better way to track intellectual property in business and music rights for artists. We have received a good number of chapters for this volume, out of which thirteen selective chapters have been considered for this book.

With blockchain technology, the businesses worldwide are adopting and building trust in every sphere. Chapter 1 entitled "A Systematic Study of Blockchain Evolution, Architecture, and Application with Use Cases" by Samanta et al. guides one to understand both non-technical and technical working principles, evolution, architecture, types, use cases, applications and tools for implementation of blockchain.

Despite the fact that blockchain technology is secured, still its wide applicability fails to gain popularity. Chapter 2 entitled "Blockchain: A Technology in Search of Legitimacy" by Rosati et al. throws light on how the players in the blockchain ecosystem attempted to build legitimacy. Their findings illustrate how the key actors engage in the blockchain and bitcoin legitimation discourse to help interpret this complex innovation and mobilize the market to adopt.

We are aware that the present world has become completely dependent on Internet. However, with intermittently connected network, the blocks can be accessed without relying on the Internet. This is well proved in the next chapter on "Using Blockchain in Intermittently Connected Network Environments" by Basu et al. They have explored possible integration of blockchain technology with intermittently connected networks towards exploiting its utility and availability in DTN environment towards improving disaster management services in the absence of end-to-end connectivity to the Internet.

However, like every coin has two sides so is the use of technology. Even though blockchain technology is said to be secure, still there are some prevailing scams till date. In this context, Chap. 4 entitled "Slaying the Crypto Dragons: Towards a CryptoSure Trust Model for Crypto-economics Blockchain versus Trust: The Expert's View of the Crypto Scammers" by Dr. Stephen Castell highlights the General Data Protection Regulation (GDPR) and its importance in the blockchain system and provides a checklist giving practical, generally applicable wording for an effective Digital Asset Disclosure exercise.

The core technology that enables the socio-technical tool, smart contracts which are the distributed ledger, is blockchain. These smart contracts require a large range of commercial applications to be taken into account. Therefore, Chap. 5 in this volume by Deval et al. on smart contracts titled as "Decentralized Governance for Smart-Contract Platform Enabling Mobile Lite Wallets Using a Proof-of-Stake Consensus Algorithm" has done a comparative analysis with Ethereum Smart Contract and found that they are not scalable for large industrial application owing to not being able to change the blockchain parameters.

Again, blockchain is not only confined to cryptocurrency. In fact, the underlying technology that supports cryptocurrency is blockchain. It prevents data feed failure and corruption. It also creates mechanism that grants protection to distributed systems. Clark et al. in their chapter titled "Blockchain Technology: Security and Privacy Issues" discussed on the use of still developing technology and also concluded that blockchain application is not possible to all applications. However, according to the author, the centralized solutions are still applicable in many cases.

As we know, privacy issues are prevailing in all industries and technology. Blockchain is a tool for keeping the data anonymously. The authors, Capraz and Ozsoy, summarized proposed models by using zero-knowledge proof which is Zero-coin and Zerocash in the chapter entitled "Personal Data Protection in Blockchain with Zero-Knowledge Proof".

With the increase in the business processes, the data protection has become a greatest challenge for the developers. Blockchain and smart contract technologies have been identified as promising approaches for supporting compliance checking and trust. The upcoming chapter by Barati and Rana entitled "Design and Verification of Privacy Patterns for Business Process Models" describes how smart contracts can be used to meet GDPR compliance verification using a number of privacy patterns for business process models.

Blockchain technology has marked its existence in almost all the sectors, and the recent one is in the field of energy. The concept of distributed renewable energy sources has grabbed attention now. The next chapter on "Blockchain Technology in Energy Field: Opportunities and Challenges" by Bai and Shen addresses the problem by proposing consensus mechanism to optimize energy transaction processing speed.

Again, energy sector is embracing the deployment of blockchain and smart contracts in a slower pace, may it be for the regulations or the complexity of the energy market. The chapter on "Blockchain Technology for Energy Transition" by Bürer et al. provides a framework for assessing the relevance and impact of blockchain technologies for enabling the energy transition from both the production and consumption sides of the equation.

Next, health care is one of the biggest sectors storing the data of the individuals. The advancements in the blockchain technology and the distributed ledger facilitate better storage of data. Chapter 11 entitled "The Feasibility and Significance of Employing Blockchain-Based Identity Solutions in Health care" by the Zhang and Kuo presents a systematic overview of the underlying motivations and principles of blockchain identities. Furthermore, they also introduced two of the popular blockchain-based identity frameworks.

Startups need funding for its survival, and initial coin offerings (ICO) are the strongest supporters of the startups now. The blockchain in health care and the relevance to raising finance with the help of ICO are trending in the market now. Thus, Chap. 12 entitled "Towards eHealth with Blockchain: Success Factors for Crowdfunding with ICOs" by Stefan Tönnissen investigates the successful initial coin offerings of eHealth startups in the field of blockchain and its impact.

Last but not least, Chap. 13 by Canbolat et al. titled "Blockchain Track and Trace System (BTTS) for Pharmaceutical Supply Chain" provide a BTTS for improving the management of pharmaceutical supply chain.

Since this volume provides a broad coverage of different aspects of blockchain technology ranging from fundamental concepts to privacy and security issues along with diverse applications, this volume can provide a strong base for naïve researchers as well as experts to explore new directions for carrying out their work further. I hope the reader gets immense benefit out of this volume.

Prof. Srikanta Patnaik

Contents

About the Editors

Prof. (Dr.) Srikanta Patnaik is Professor in the Department of Computer Science and Engineering, Faculty of Engineering and Technology, SOA University, Bhubaneswar, India. He has received his Ph.D. (Engineering) on Computational Intelligence from Jadavpur University, India, in 1999 and supervised 25 Ph.D. theses and more than 50 Master theses in the area of Computational Intelligence, Soft Computing Applications and Re-Engineering. Dr. Patnaik has published around 100 research papers in international journals and conference proceedings. He is the author of 2 textbooks and 35 edited volumes and few invited book chapters, published by leading international publisher like Springer-Verlag and Kluwer Academic. Dr. Patnaik was Principal Investigator of AICTE sponsored TAPTEC project "Building Cognition for Intelligent Robot" and UGC sponsored Major Research Project "Machine Learning and Perception using Cognition Methods." He is Editor-in-Chief of *International Journal of Information and Communication Technology* and *International Journal of Computational Vision and Robotics* and also Associate Editor of *International Journal of Granular Computing, Rough Sets and Intelligent Systems* (IJGCRSIS) and *International Journal of Telemedicine and Clinical Practices* (IJTMCP), published from Inderscience Publishing House, England. Dr. Patnaik is also Editor of *Journal of Information and Communication Convergence Engineering*, published from Korean Institute of Information and Communication Engineering. He is also Editor-in-Chief of Book Series on "Modeling and Optimization in Science and Technology" published from Springer, Germany.

Prof. Tao-Sheng Wang is Doctor, Professor of Applied Economics (International Trade Direction), Specialist of General Office of the State Council, Director of Chinese Society of World Economy, Vice Chairman of the Expert Committee of China Association of Trade in Service and Master Tutor. He is Dean of the School of Business of Hunan International Economics University and Chief Expert of the provincial-level social science key research base. He is also a foregoer of provincial-level key construction course "International Trade," provincial-level teaching team "International Trade" and provincial-level characteristic specialty "International Economics and Trade." His main research areas are international economic and trade theory and policy; institutional innovation and international trade; international competitive advantage and competitive strategy.

Prof. Tao Shen is the Deputy Dean of the College of Information Engineering and Automation, Kunming University of Science and Technology. He received his Master and Ph.D. degree from Illinois Institute of Technology in USA and Bachelor's degree from the University of Electronic Science and Technology of China. He has conducted and participated in over 20 research projects, including projects supported by the National Natural Science Foundation of China, Key Project of Applied Basic Research Programs of Yunnan Province. He has published more than 60 papers indexed by SCI/EI, 1 book and received 11 national patents. His main research fields including intelligent detection and sensing, artificial intelligent, blockchain and energy internet.

Prof. (Dr.) Sushanta Kumar Panigrahi has received his Ph.D. (Engineering and Information Technology) on Artificial Intelligence and Soft Computing from Fakir Mohan University, Balasore, Odisha, India, in 2016, and supervised more than 25 Master theses in the area of Computational Intelligence, Soft Computing Applications and Re-Engineering. Dr. Panigrahi is well acclaimed for his stints in teaching at various institutions. His academic career spans over more than one decade in which he has earned the reputation of an admirable teacher in areas of Optimization, Database Management, Soft Computing and Data Science. Dr. Panigrahi has published more than 38 research papers in international journals and conference proceedings. He has attended and presented papers in a number of IEEE, Elsevier and Springer conferences. Dr. Panigrahi is also an active reviewer of a number of journals of IEEE, Elsevier, Inderscience and Springer in the discipline of Computer Science and Engineering.

Chapter 1
Introduction to Blockchain Evolution, Architecture and Application with Use Cases

Sidharth Samanta, Bhabendu Kumar Mohanta, Deepti Patnaik, and Srikanta Patnaik

Abstract Blockchain is not a hype, it is here to stay. Before you dive deeper into the business applications of the Blockchain, you have to understand the fundamentals and the functionalities of the Blockchain. This chapter will guide you to understand both non-technical and technical working principles, evolution, architecture, types, use cases, applications, and tools for implementation of Blockchain. Although there is no technical prerequisite of the chapter, a basic knowledge of network systems and cryptography is required for better understanding.

1.1 Layman's Introduction to Blockchain

Suppose you are lending some money to your friend and both of you agreed on some terms like the period and the interest rate. To overcome the worst-case scenarios like denial and false claim, you pick another person as a witness of the transaction and the agreement. Every transaction has to be gone through that person's approval. In this process, no individual can claim or deny any transaction.

As you are thinking about the worst-case scenario, humans are prone to circumstances and manipulations such as unavailability, foul play, and partiality. To over-

S. Samanta
Department of CSA, Utkal University, Bhubaneswar, India
e-mail: samantasidharth@gmail.com

B. K. Mohanta (✉)
Department of CSE, IIIT, Bhubaneswar, India
e-mail: bhabendukumar@gmail.com

D. Patnaik
Department of Finance, IIMT, Bhubaneswar, India
e-mail: deeptipatnaik5@gmail.com

S. Patnaik
Department of CSE, SOA University, Bhubaneswar, India
e-mail: patnaik_srikanta@yahoo.co.in

© The Author(s), under exclusive license to Springer Nature Singapore Pte Ltd. 2021
S. Patnaik et al. (eds.), *Blockchain Technology and Innovations in Business Processes*,
Smart Innovation, Systems and Technologies 219,
https://doi.org/10.1007/978-981-33-6470-7_1

come this, you and your friend assign many people as witnesses, where everyone maintains the transnational records, and the majority consensus is required to verify a claim.

In this way, neither you nor your friend can claim or deny any transaction. Because unlike an individual, it is very difficult to manipulate a group as a whole. The Blockchain is based on this principle where all the witnesses are called nodes that are connected in a distributed network. All nodes maintain a digital ledger consisting of transactions bundled in batches. These batches are called blocks, and every block is connected with previous ones in a chained manner, so it is called the Blockchain.

1.2   The Blockchain Technology

Ever since its inception, the Blockchain serves as the public transaction ledger. It is a timestamped series of immutable records, which stores a large amount of data and it is managed by groups of computers that are not owned by any single entity. Every block of data is bound and secure using cryptographic principles. As the network is not owned or controlled by a central authority, the information in this network is open for every node in the network.

Blockchain technology is a simple and innovative way of forwarding information from one part of the world to others in a safe and automated way. It starts with generating a block of transactions. A block needs to be confirmed by the majority of computers (nodes) scattered around the internet by a consensus mechanism. After verification, the block is attached to the existing chain, and the chain is stored in the digital ledger of every node. In this way, it will create a unique and immutable record as shown in the top section of Fig. 1.1. Some key features of this technology are decentralization, transparency, and immutability.

Blockchain is built on the top of peer-to-peer topology. It is a distributed ledger technology (DLT) that allows data to be stored globally, on thousands of servers. With this technology, each page in a ledger of transactions forms a block. That block has an impact on the next block or page through cryptographic hashing. Figure 1.1 shows how blocks are created and transactions are approved in a Blockchain. We will further discuss the process in a systematic layered manner in this chapter.

1.3   Types of Blockchain

With respect to nature and the platform, there are three kinds of Blockchain architectures such as public, consortium, and private Blockchain. A public Blockchain is open to all, but a private Blockchain can only be assigned by some selected individuals. A consortium Blockchain is a hybrid between public and private Blockchain, where a group of private Blockchains is combined. Although the consortium Blockchain is quite similar to the private Blockchains, Table 1.1 shows the detailed differences and similarities between the three types of Blockchains.

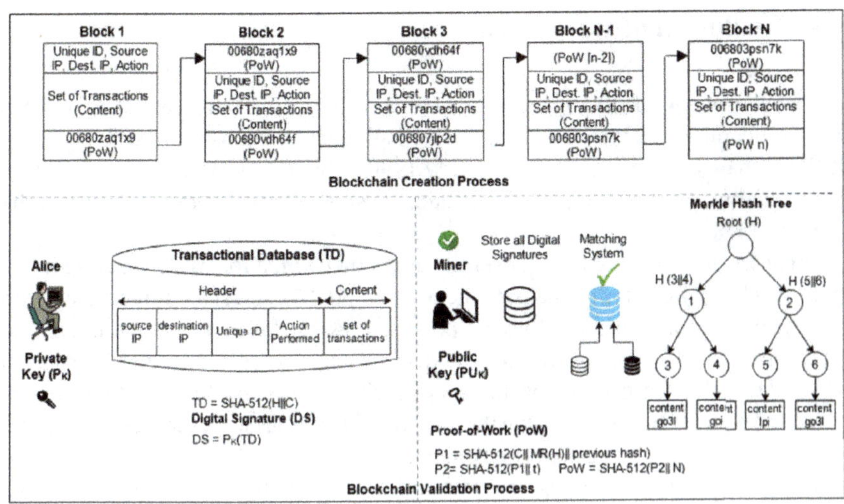

Fig. 1.1 Block creation and validation process [1]

Table 1.1 Comparison among different types of Blockchains [2]

Properties	Public	Consortium	Private
Nature	Open and decentralized	Controlled and restricted	Controlled and restricted
Participants	Anonymous and resilient	Identified and trusted	Identified and trusted
Consensus procedures	PoW, PoS, DPoS	PBFT	PBFT, RAFT
Read/Write permission	Permissionless	Permissioned	Permissioned
Immutability	Infeasible to tamper	Could be tampered	Controlled and could be tampered
Efficiency	Low	High	High
Scalability	High	Low	High
Transaction approval frequency	Long (10 min or more)	Short	Short
Energy consumption	High	Low	Low
Transparency	Low	High	High
Observation	Disruptive in terms of disintermediation	Cost effective due to less data redundancy and higher transaction times	Cost effective due to less data redundancy and higher transaction times
Example	Bitcoin, Ethereum, Litercoin, Factom, Blockstream, Dash	Ripple, R3, Hyperledger	Multichain, Blockstack, Bankchain

1.4 Evolution of Blockchain

Blockchain is one of the biggest innovations of the 21st century affecting various sectors, i.e, financial, manufacturing, as well as educational institutions. The history of Blockchain has emerged from the early 1990s which is unknown to many. Before understanding the evolution of Blockchain, we should focus on the evolution of the internet.

What we are using today is conventional internet which focuses on sharing information, All the websites are helping the users to publish and share information with other people on the internet. All enterprises are using this conventional internet to share information with stakeholders and gives them good user experience, from which they are getting more revenue out of their business. At the same time, we are trying to create an Internet of value that carries a value transacted through the internet which is only possible through Blockchain.

In the early 1990s, many people started focusing on digital currencies. In 1982, David proposed digital cash through a paper with a blank signature for untraceable payments, So many people got involved in creating digital currencies.

In 2008, Satoshi Nakamoto [3] conceptualized bitcoin in his paper and launched it in the next year. The researchers found that his concept was very helpful in solving many enterprise problems. So less than a decade, Blockchain evolved so rapidly from version 1.0 to 3.0.

After the launch of Blockchain in 2008, Ripple another notable currency by Ryan Fugger, strengthen the Blockchain version 1.0 then Vitalik Buterin creator of Ethereum started the journey of Blockchain 2.0 [4] by bringing the Blockchain into enterprise and Linux Foundation joined their Hyperledger project, and after them, Cardano Foundation and IOTA took it to the next level to version 3.0. Let us understand them to stage by stage.

1.4.1 Blockchain 1.0: Programmable Money and Cryptocurrencies

- It mainly focuses on cryptocurrencies and their values.
- A cryptocurrency is a digital asset designed to work as a medium of exchange that uses strong cryptography to secure financial transactions, control the creation of additional units, and verify the transfer of assets. They use de-centralized control as opposed to centralized digital currency and central banking systems.

1.4.2 Blockchain 2.0: Smart Contracts

The smart contract is a self-executable program. The use cases of smart contracts are explained in detail in [5].

- It mainly focuses on enterprises using smart contracts.
- Smart Contracts are autonomous computer programs using Blockchain technology that once started, execute automatically and mandatory the conditions defined beforehand, such as facilitation, verification, or enforcement of the negotiation or performance of a contract.
- Small contracts are triggered automatically only when required conditions are satisfied which eliminates the need for trusted third parties.

1.4.3 Blockchain 3.0: Decentralised Web

- Almost all internet user's data are handled by The Big 5 i.e., Google, Microsoft, Apple, Amazon, and Facebook. They can monetize the data without permission to share with the Govt. Agencies or other agencies without their knowledge.
- Also, these collected data are vulnerable to hacking, the user may lose entire data.
- Current centralized architecture of the Internet created many issues, one major issue is that if the server goes down the data becomes inaccessible, and also the centralized architecture is susceptible to DDOS attacks [6]. These decentralized content delivery networks using Blockchain technology can guard against DDOS attacks.
- The decentralized application is run by many users on a decentralized network with trustless protocols. They are designed to avoid any single point of failure.

1.5 Ledger System for Blockchain

The Blockchain is a ledger system, which maintains the transnational records. The ledger systems can be of many types for managing authority, such as the centralized system, the decentralized system, and the distributed system. In this section, we will discuss the centralized system, which is the current date industry standard and the distributed systems.

1.5.1 Centralized Ledger Systems

As per the nomenclature, a system having centralized control with all administrative authority is called a centralized ledger system (see Fig. 1.2). All systems available

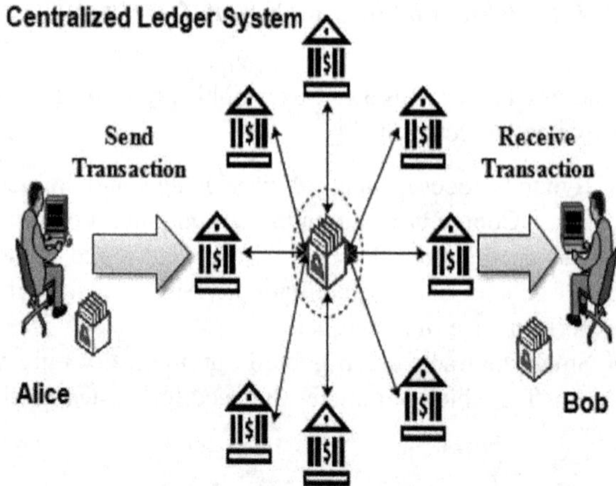

Fig. 1.2 Centralized ledger system [1]

in the network depends on a single system, which authenticates and controls tasks. These systems are simple and pretty straightforward to design, maintain, and govern. The conventional banking system works in this principle, where every user needs the bank's permission to perform transactions. But they inherit some limitations such as:

- Central points of failure can lead to reliability concerns.
- Vulnerability to attack can lead to security concerns.
- The centralization of power can lead to unethical operations.
- Difficult and expensive to scale.

1.5.2 Distributed Ledger Systems

In contrast to the centralized system, a distributed ledger system does not have a central authority. As per the nomenclature, a ledger system with no centralized control and every node has equal authority and control to maintain the ledger is called a distributed ledger system (see figure 1.3). In this system, every node maintains a local ledger. These systems are difficult to design, maintain, and govern, but pretty easy to scale as compared to centralized systems. The distributed ledger system possesses the following advantages over the centralized system:

- Absence of central point failure results in superior reliability.
- Attack resistance makes is more secure.
- The symmetric system with equal authority.
- High scalability.

The Blockchain is built on the top of distributed ledger systems because of its above-mentioned advantages. Scalability and distributed authority gives the advantage of security and helps the transactions be reliable. As each node maintains a local ledger of transactions, this system is transparent.

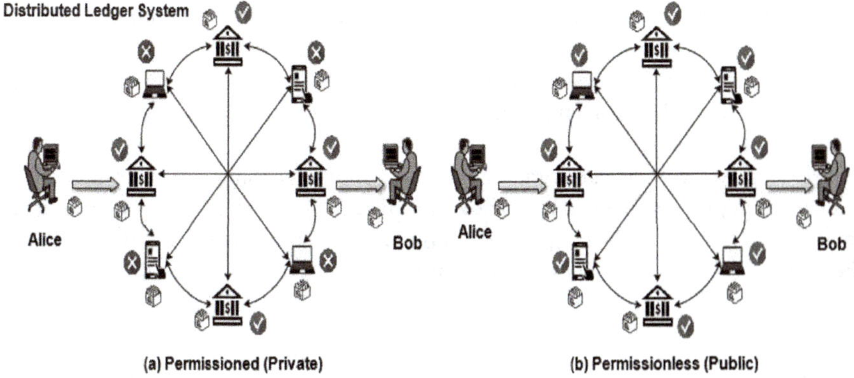

Fig. 1.3 Distributed ledger system [1]

1.6 Layers of Blockchain

As of writing, Blockchain is deployed in many areas with very distinct applications. Unlike the internet, Blockchain's structure can not be generalized as different use cases will require different structures. So, this section will discuss layers of a generic Blockchain architecture. There are five layers of a typical Blockchain, listed in Figure 1.4.

1.6.1 Hardware or Infrastructure Layer

The first layer of the Blockchain is the infrastructure layer or hardware layer. The Blockchain contents are hosted on a server or a high-performance computer situated

Fig. 1.4 Layered architecture of Blockchain

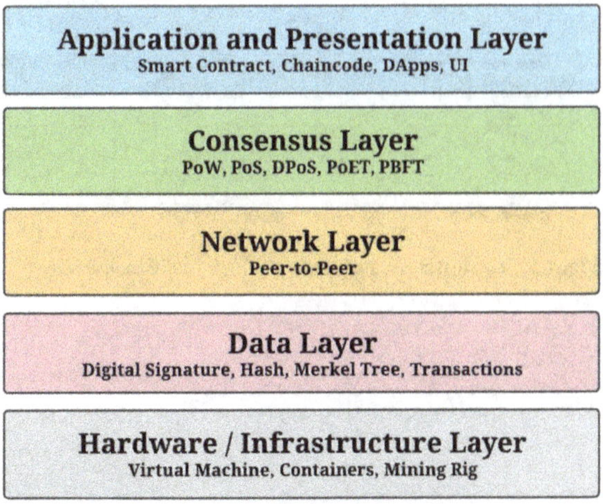

remotely referred to as nodes. Each node is responsible for validating transactions, organizing the transactions inside the block, broadcasting in the Blockchain network, etc. These nodes are also responsible for adding a block into the chain after consensus. Node is considered a virtual machine that consists of virtual resources such as storage, network and servers.

Physically, the nodes are computer having a very high computational capability. It can be a computer connected to a rig of multiple GPUs called a mining rig or a virtual machine in the cloud. When a device is connected to the network of a Blockchain, it becomes a node. These nodes are decentralized as well as distributed.

1.6.2 Data Layer

In the Blockchain, an individual block is a container data structure that contains many more things than a list of transactions. This layer is responsible for the structure of the block in the Blockchain. In bitcoin, a typical block contains roughly 500 transactions on an average, which is about 8 MB in size. In the Blockchain, an individual block is a container data structure that contains many more things than a list of transactions. This layer is responsible for the structure of the block in the Blockchain. In bitcoin, a typical block contains roughly 500 transactions on an average, which is about 8 MB in size. The data items in a single block are divided into two sections such as block header and block body. The Block header contains the metadata about the particular block, whereas the block body contains the list of transactions. Figure 1.5 shows the inner con taints of the blocks.

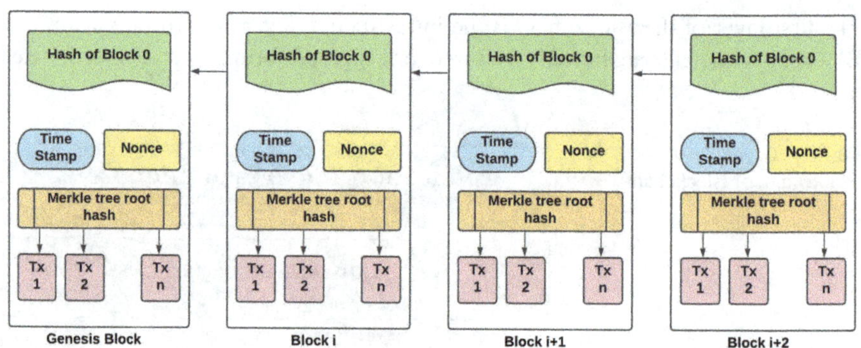

Fig. 1.5 The different attributes of block in a Blockchain

1.6.2.1   Block Header

The Block header section of the block contains all the data about the data in the particular block. The block header of the bitcoin block is further subdivided into six subsections, such as:

- Version: The version number of the software.
- Last Block: The hash value of the previous block.
- Merkle Root: The root of the Merkle tree.
- Time: The time in second since 1970-01-01 T00:00UTC.
- Miner: Information about the miner.
- Target: The hash size in bits.
- The Nonce: A variable incremented by the proof of work.

1.6.2.2   Block Body

The block body holds all the confirmed transactions in the block. Whenever a transaction requires to be validated, the nodes present in the network evaluates the transactions present in the block body. In the bitcoin transaction, it contains the attributes like the Hash value of the transaction, Sender's address, Receiver's address, Amount, Fees, Time of a transaction, Number of confirming nodes, etc.

1.6.3   Network Layer

In the Blockchain, the network layer is also known as the P2P or peer-to-peer layer. A peer-to-peer network is a kind of computer network, where the nodes are distributed in a network and share the workload. In the Blockchain, there are two kinds of nodes, such as the full node and the light node. Full nodes are responsible for the verification and validation of transactions, mining, and consensus. However, the light nodes only keep the Blockchain headers and send it whenever required. The network layer is responsible for communication between nodes. This layer takes care of the discovery of nodes, transactions, and backpropagation. In some cases, this layer is also referred to as the propagation layer. This layer ensures the communication between nodes for the synchronization of a valid chain of blocks.

1.6.4   Consensus Layer

For the Blockchain, the consensus protocol is the core. This layer is the most crucial and critical layer for any Blockchain. This layer is responsible to validate the blocks,

Table 1.2 Comparision of consensus algorithms [8]

	PoW	PoS	PBFT
Type of blockchain	Permissionless	Permissionless and Permissioned	Permissioned
Transaction Finality	Probabilistic	Probabilistic	Deterministic
Token	Yes	Yes	No
Example	Bitcoin, Ethereum	Ethereum	Hyperledger fabric

ordering the block. The consensus mechanism [7] is the reason that Blockchains are more secure and trusted than any other conventional method.

The consensus mechanism varies according to the application area of the Blockchain. For example, the consensus for a permissionless Blockchain network such as bitcoin Ethereum is known as probabilistic consensus. This type of consensus mechanism guarantees the consistency of the ledger. Although there is a possibility of some variations like some participants have a different view of the block. In other cases, permission Blockchains like Hyperledger Fabric follow a deterministic algorithm. In these kinds of Blockchain networks, there are some ordering nodes to validate the block order sequence, and this sequence is considered as final and true. This results in a zero probability of fork. Table 1.2 outlines a shallow comparison of consensus algorithms such as PoW (Proof of Work), PoS (Proof of Stake), and PBFT (Practical Byzantine Fault Tolerance).

1.6.5 Application Layer

In the Blockchain, the application layer or the presentation layer is the bottom-most layer. This layer is comprised of smart contracts, chain code, and apps. The application layer is further divided into two sub-layers, such as the application sub-layer and execution sub-layer.

The application sub-layer is a collection of software packages that are used by the users to interact with the Blockchain. This package comprises scripts, APIs, UI, and frameworks. The user interfaces of the application sub-layer use APIs to interact with the Blockchain network. The execution sub-layer constitutes smart contracts, chain code, and underlying rules. This sub-layer has the actual executable code and rules to be followed.

A transaction propagates from the application sub-layer to the execution sub-layer, where the rules of the smart contract validate it. The application sub-layer sends the instructions to the execution sub-layer, which performs the execution to ensure the deterministic nature of the Blockchain.

1.6.5.1 Smart Contract

A smart contract is a code with business logic, which is further identified by a unique address. It contains functions that are executed when a transaction takes place. Depending on the rules mentioned in the smart contract, the transaction is validated.

1.6.5.2 Chaincode

Smart contracts are the transaction rules that control the life cycle of transaction instances. In Blockchain networks, the smart contracts are packed together into chain codes. A chance can contain many smart contracts. Considering the example of an insurance chain code, it can contain smart contracts for liability, claim, processing, etc.

The chain code defines the schema of the ledger initiates it and updates it by responding to the queries. It can release its events to allow other applications to follow the chaincode events, which can lead to initiate downstream functions or processes.

1.6.5.3 DApps

dApps in the end, Blockchains like Ethereum, bitcoin are distributed ledger system that needs a distributed application to work as a whole. So it needs a special kind of software named as DApps. It stands for decentralized applications, used to leverage smart contracts or chain code. It is a web application that interacts with the chain code or smart contract. The DApps are user friendly and not controlled by any organizations.

1.7 Use Cases and Applications of Blockchain

Blockchain technology has a wide application area as shown in Fig. 1.6. With the use of Information and Communication Technologies (ICTs) and IoT, most of the applications are converted into smart applications from traditional architecture. Different application areas like Healthcare, Financial, IoT, legal perspective, Government, Intelligent transportation system, Power Grid, E-business, Commercial world, and Reputation system are explained in detail [9]. Blockchain help in design secure and trustful among the user participants in the applications. All the transactions are stored in digital ledger format and it is available to every member of the network. The full Blockchain can be stored by the high-end system. If the system is lightweight like a mobile phone, tablet, then users can use the InterPlanetary File System (IPFS) concepts to store the Blockchain reference. Architecture point of view decentralized or

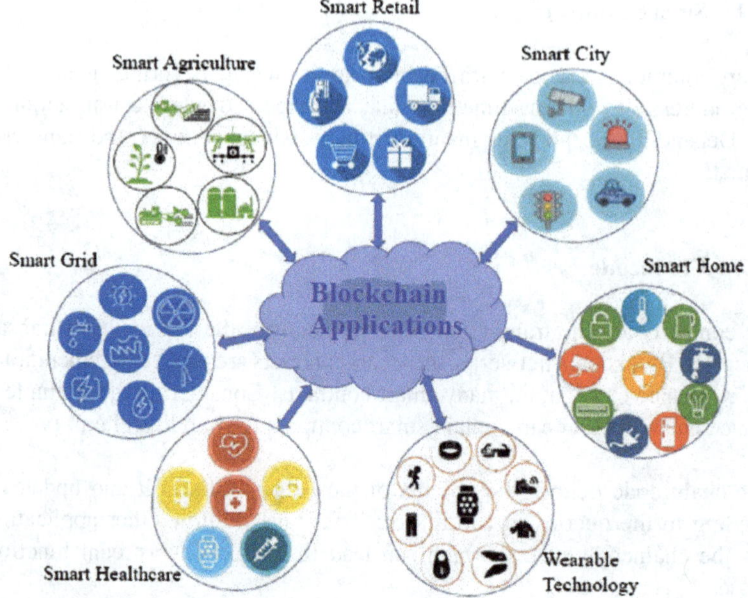

Fig. 1.6 Different emerging application area of Blockchain

distributed network is developed to implement an application that will eliminate the network failure like a single-point failure. Using Blockchain's real-time monitoring of the different attributes value changes can be identified. Each transaction can be verified and validated using past transactions.

1.8   Development Tools for Blockchain

With the adoption of Blockchain, the options for its development and deployment is increasing at an exponential rate. As of writing, there are many dozens of tools that are available to develop Blockchains and smart contract frameworks. Some most popular tools are mentioned in Table 1.3. The table also contains some brief information about the particular tool and its official source. Table 1.4 gives a piece of detailed information about some popular Blockchain platforms. In paper [10, 11] authors proposed Blockchain authentication protocols for IoT devices. IoT is the most emerging technology, and it is being used in most of the application areas. But the IoT system has some security and privacy issues which need to be addressed, so that its end-users build trust to use the applications. Blockchain integration with IoT solved some of the security issues.

Table 1.3 List of blockchain tools

Tool	Brief description
MetaMask	It is a browser extension that allows you to browse Ethereum Blockchain enabled websites. https://metamask.io/
web3.js	This library is the Ethereum compactable JavaScipt API which implements the Generic JSON RPC spec to interact with a local or remote Ethereum nodes. https://github.com/ethereum/web3.js/
Truffle Framework	The most popular Ethereum development framework and asset pipeline for Ethereum. https://truffleframework.com/
pyethereum	It is the Python core library for the Ethereum project https://github.com/ethereum/pyethereum/
web3j	It is a light weight, reactive, and typesafe Java and Android library to use with smart contracts and integrate with clients on the Ethereum Network. https://github.com/web3j/web3j
Go Ethereum	It is also known as "Geth", can be used as a standalone client or a library that you can embed in Go, Android or iOS projects. https://ethereum.github.io/go-ethereum/
Solidity	It is a contract-oriented, high-level language for implementing smart contracts. It was influenced by C++, Python and JavaScript, http://solidity.readthedocs.io/
Ganache CLI	It is a part of Truffle's suite of dev tools which uses EthereeumJs Is to simulate client behaviour to make developing Ethereum applications faster and easier. https://truffleframework.com/ganache/
Mythril	It is a security analysis engineer and platform used to analyze the Ethereum smart contracts and Dapps. It integrates with several commonly used IDEs. https://consensys.net/diligence/mythril.html

1.9 Summary

In this chapter, we covered the concept of Blockchain, the evolution, its types, the architecture, use cases, applications, and tools to implement Blockchain. In summary, the Blockchain is more than a platform of cryptography or some fancy terminology. Although the structure and the parameters involved in the Blockchain changes with the application areas, this chapter gives the most generalized form of Blockchain.

The efficiency, security, and the trust provided by Blockchain technology, and its adoption in the financial community will result in rapid growth in research and implementation in different business communities. In the era of globalization, the Blockchain will help to maintain trust and integrity. This chapter will conclude with brief descriptions of the contribution of Blockchain's game-changing solutions in different businesses, which are explained thoroughly in further chapters.

Table 1.4 List of blockchain platforms [1]

Platform	Network type	Purpose	Prog. language	Crypto-currency	Consesus mechanism	Smart contract enabled	Transaction cost for timing	Applications
Ethereum	Public/private permissionless	B2C businesses	Smart contract code written in solidity	In-built (Ether)	PoW (PoS- in future)	Yes	Yes (high)	banking, commodity trade finance, supply chain management, insurance, energy grid, oil & Gas, real estate
R3Corda	Private, permissioned	B2B businesses	Smart Contract code written in Kotlin, Java	not in-built	Pluggable (voting process, BFT)	Yes	No	Banking, financial services
Hyperledger fabric	Private, permissioned	P2P, B2C operations	Chaincode written in GoLang, Java	Not in-built (can be modeled in chaincode)	Pluggable (PBFT)	Yes	No	Supply chain for pharmaceuticals, trade financing, smart energy, supply chain management
MultiChain	Private, permissioned	B2B operations	Python, C, JavaScript, PHP, Ruby	Not in-built	PBFT	No	No	Financial transactions, e-commerce
HydraChain	Private, permissioned	B2B operations	Python based smart contracts	Not in-built	BFT	Yes	No	Financial services
Openchain	Private, permissioned	P2P, B2B operations	Javascript	Not in-built	Partitioned consensus	Yes	No	Digital asset management
IBM Blockchain	Private, permissioned	B2B operations	GoLang, Javascript	Not in-built	Pluggable (PoW)	Yes	Yes (low)	Healthcare payments, trade and supply chain finance

(continued)

Table 1.4 (continued)

Platform	Network type	Purpose	Prog. language	Crypto-currency	Consesus mechanism	Smart contract enabled	Transaction cost for timing	Applications
IOTA	Public	B2B operations	Python, C, JavaScript	Not in-built	PoW, PoS	No	No	Financial, telecommunication, intelligent energy, e-healthcare
Bitcoin	Public/private permissionless	B2B, B2C operations	C++	In-built (bitcoin)	PoW	Yes	Yes (high)	Government, financial audit trails
Litecoin	Public/private permissionless	B2B, B2C operations	C++	In-built (litecoin or LTC)	PoW	Yes	Yes (low)	Banking, financial services
BigchainDB	Public/private, permissionless	B2C operations	SQL, NoSQL	No-inbuilt	BFT, federation with voting permissions	Yes	No	Intellectual property, human resources, identity verification, supply chain, land registry
Chain core	Private, permissionless	B2B operations	Java, Ruby, Node, JS	Not in-built	Federate consensus	Yes	No	Financial assets Eris:db (eris:db, 2016) private, permissioned B2B operations Javascript not in-built BFT yes yes (low) banking services
Quorum	Private, permissionless	B2B operations	Smar contract written in solidity	In-built (JPM Coin)	Majority voting, on-demand creation	No	No	Banking, financial, insurance services
Stellur	Public/private, permissionless	B2B operations	C++, Java, Python, Golang, Javascript	In-built (lumen)	Stellar consensus protocol	Yes	Yes (low)	Telecommunication sector, banking, financial services

References

1. Aggarwal, S., Chaudhary, R., Aujla, G.S., Kumar, N., Choo, K.K.R., Zomaya, A.Y.: Blockchain for smart communities: Applications, challenges and opportunities. J. Network Comput. Appl. **144**, 13–48 (2019)
2. Monrat, A.A., Schelen, O., Andersson, K.: A survey of blockchain from the perspectives of applications, challenges, and opportunities. IEEE Access **7**, 117134–117151 (2019)
3. Nakamoto, S.: Bitcoin whitepaper (2008). https://bitcoin.org/bitcoin.pdf. 17 July 2019
4. Sun, S.F., Au, M.H., Liu, J.K., Yuen, T.H.: RINGCT 2.0: A compact accumulator-based (linkable ring signature) protocol for blockchain cryptocurrency Monero. In: European Symposium on Research in Computer Security, pp. 456–474. Springer, Cham (2017, September)
5. Mohanta, B.K., Panda, S.S., Jena, D.: An overview of smart contract and use cases in Blockchain technology. In: 2018 9th International Conference on Computing, Communication and Networking Technologies (ICCCNT), pp. 1–4. IEEE, New York (2018, July)
6. Mirkovic, J., Reiher, P.: A taxonomy of DDoS attack and DDoS defense mechanisms. ACM SIGCOMM Comput. Commun. Rev. **34**(2), 39–53 (2004)
7. Panda, S.S., Mohanta, B.K., Satapathy, U., Jena, D., Gountia, D., Patra, T.K.: Study of blockchain-based decentralized consensus algorithms. In: TENCON 2019-2019 IEEE Region 10 Conference (TENCON), pp. 908–913. IEEE, New York (2019, October)
8. Acharya, V., Yerrapati, A.E., Prakash, N.: Oracle Blockchain Quick Start Guide: A Practical Approach to Implementing Blockchain in Your Enterprise. Packt Publishing Ltd. (2019)
9. Mohanta, B.K., Jena, D., Panda, S.S., Sobhanayak, S.: Blockchain technology: A survey on applications and security privacy challenges. Int. Things **8**, 100107 (2019)
10. Hammi, M.T., Hammi, B., Bellot, P., Serhrouchni, A.: Bubbles of trust: A decentralized blockchain-based authentication system for IoT. Comput. Security **78**, 126–142 (2018)
11. Mohanta, B.K., Sahoo, A., Patel, S., Panda, S.S., Jena, D., Gountia, D.: DecAuth: decentralized authentication scheme for IoT devices using the Ethereum blockchain. In: TENCON 2019-2019 IEEE Region 10 Conference (TENCON), pp. 558–563. IEEE, New York (2019, October)

Chapter 2
Blockchain: A Technology in Search of Legitimacy

Pierangelo Rosati, Theo Lynn, and Grace Fox

Abstract Blockchain can potentially disrupt business processes across many industries. Despite substantial hype, blockchain adoption still remains quite low. Research suggests that the success and adoption of a new technology depends on whether it is deemed legitimate by key stakeholders. This is extremely important in the context of blockchain given the controversial reputation that some derivative innovations, such as Bitcoin, have gained over time. This chapter explores and compares how four types of business actors which play a prominent role in the blockchain ecosystem—the media, IT, financial services and consulting firms—attempted to build legitimacy around blockchain and Bitcoin on Twitter. Findings suggest that the messaging strategies employed by firms for blockchain and Bitcoin differ. While no dominant firm-level messaging strategies to build legitimacy around Bitcoin emerge, firms primarily employ three micro-level legitimation strategies to build legitimacy around blockchain. More specifically, communication strategies around blockchain focus on publishing the involvement of influential actors (pragmatic legitimacy), highlighting blockchain ongoing developments and associated market responses (cognitive legitimacy), and presenting technological advancements directly or indirectly related to blockchain (cognitive legitimacy). The findings illustrate how the key actors engage in the blockchain and Bitcoin legitimation discourse to help interpret this complex innovation and mobilise the market to adopt.

P. Rosati (✉) · T. Lynn · G. Fox
Irish Institute of Digital Business, Dublin City University, Dublin, Ireland
e-mail: pierangelo.rosati@dcu.ie

T. Lynn
e-mail: theo.lynn@dcu.ie

G. Fox
e-mail: grace.fox@dcu.ie

S. Patnaik et al. (eds.), *Blockchain Technology and Innovations in Business Processes*,
Smart Innovation, Systems and Technologies 219,
https://doi.org/10.1007/978-981-33-6470-7_2

2.1 Introduction

The increasing adoption of digital technology is rapidly changing business processes and business models. Blockchain technology is often regarded as one of the most promising IT innovations in recent times [1] and has attracted considerable attention from the media, practitioners, academics and policymakers [2]. Beck [3] defines blockchain as *"a tamper-resistant database of transactions consistent across a large number of nodes and is cryptographically secured against retrospective manipulations, and it uses a consensus mechanism to keep the database consistent whenever new transactions need to be validated"*. However, blockchain is a complex technology which has also philosophical, cultural and ideological implications [4]. As such, Mougayar [4] suggests to combine the more technical definitions of blockchain with a business and a legal definition. For Mougayar, depending on whether one approaches blockchain from a business or legal perspective, it can be defined as an *"exchange network for moving value between peers"* or *"a transaction validation mechanism, not requiring intermediary assistance"* [4].

Blockchain technology was originally proposed in 2008 by Satoshi Nakamoto, with the intent of enabling *"an electronic payment system based on cryptographic proof instead of trust, allowing any two willing parties to transact directly with each other without the need for a trusted third party"* [5]. This payment system was named *Bitcoin* and is, undeniably, the most well-known application of blockchain so far. Distrust in financial institutions following the global financial crisis and the increasing calls for more efficient payment systems were the main drivers behind the inception of Bitcoin [6–8]. Bitcoin is also referred to as a *cryptocurrency* as the system leverages cryptographic algorithms to create irreversible proof of transactions [5, 9, 10]. The volume of Bitcoin transactions and its market capitalisation has grown dramatically over the last decade together with the number of businesses accepting Bitcoin payments [11–13]. Blockchain has arguably benefitted from the success and the popularity of Bitcoin. However, this increased awareness has come with some significant drawbacks due to the questionable reputation that Bitcoin, and cryptocurrencies more generally, has developed over time. In fact, while the original objective of Bitcoin was to eradicate payment fraud and transaction costs, the anonymity offered by the system has made Bitcoin particularly attractive for actors involved in unethical or fraudulent activities [13–15].

Extant research suggests that creating legitimacy around IT innovations is a stepping stone towards widespread adoption [16–19]. This is particularly relevant for technologies like blockchain that can radically change business processes and whose potential applications, benefits and limitations are still being explored. The success of a nascent technology such as blockchain is arguably dependent on attracting interest from a wide range of organisations [20]. In this chapter, we investigate and contrast four types of key actors in the blockchain and Bitcoin ecosystem (i.e. media, IT, financial services and consulting firms) leveraged social media to legitimise these two innovations over a 12-month period. More specifically, we leverage the concept of *organising visions* proposed by Swanson and Ramiller [19] and the IT Legitimation

Taxonomy proposed by Kaganer et al. [16] to explore if and how a common conceptualisation of blockchain and Bitcoin, and their functional attributes, are formed through generative discourse within a community.

This chapter is organised as follows. Next, we briefly summarise the theoretical background underpinning IT legitimation and adoption. In Sect. 2.3, we present the empirical context and the research methodology. This is followed by the results of our analysis and by a discussion of our findings, the practical implications and limitations of our research, and potential areas for future research.

2.2 Theoretical Background

Technology adoption is a topic that has been widely discussed in the information systems literature leading to the creation of a number of different theoretical frameworks from Diffusion of Innovation theory (DOI) [21] to Technology Acceptance Model (TAM) [22], as well as integrative theories such as the Unified Theory of Acceptance and Use of Technology (UTAUT) [23] and Human–Organisation–Technology fit (HOT-fit) [24]. All these frameworks have been applied to investigate the determinants of technology adoption across different contexts and industries, but the general tendency has been to focus on different characteristics of the technology, of the adopting organisation [25], or on the characteristics and perceptions of the final user. This body of literature has provided valuable insights into the dynamics of technology adoption as summarised in various literature reviews [23, 26–29]. Technology adoption models are relevant when exploring adoption decisions at the individual level. However, in decentralised ecosystems such as blockchain, there is a need to explore these dynamics from a community and inter-organisational perspective. At early stages of diffusion, the success of a technology innovation, in this case blockchain, is heavily reliant on generating positive interest from a diverse array of organisations [20]. In the case of blockchain, generating interest from related communities and mobilising this interest is key.

Legitimation research seeks to understand the rationale behind decisions to adopt or reject IT innovations [30]. In line with Suchman [31], we view legitimacy as "[...] *a generalised perception or assumption that the actions of an entity are desirable, proper, or appropriate within some socially constructed systems of norms, beliefs, and definitions*". Within this definition, we consider four types of legitimacy:

- Cognitive legitimacy—the diffusion of knowledge within the community about the innovation [32]. This is achieved through comprehensibility and taken-for-grantedness [31];
- Pragmatic legitimacy—the self-interest and utility to the organisation's stakeholders [31];
- Normative legitimacy—the alignment of the innovation with moral rightness [31];
- Regulative legitimacy—the alignment of the innovation with legal best practice [16].

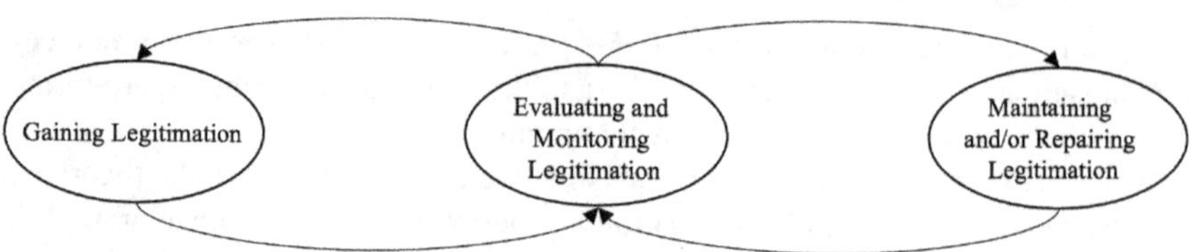

Fig. 2.1 Integrated legitimation activity model [35]

In order to explore legitimation efforts within a community, Flynn and Hussain [33] developed the Legitimation Activity Model (LAM), a seven-staged cyclical model to explain the relation between legitimation seekers and providers. Although the LAM was designed as an organisation-level framework, it can also be applied to include a range of stakeholders [34]. LAM was further developed to form the Integrated Legitimation Activity Model (ILAM) (Fig. 2.1). ILAM recognises the need for continuously monitoring and evaluating the outcomes of different legitimation efforts and adjust the corresponding strategies as a result [35]. This study builds such a need and aims to (i) understand the strategies utilised by organisations to legitimise blockchain and Bitcoin and (ii) provide practical recommendations to inform legitimation strategies adopted among these communities.

This chapter explores the discourse within the Blockchain community on Twitter adopting the lens of organising visions, according to which the social system both shapes and employs a vision of a new innovation which becomes crucial to its success [36]. Organising visions within a community help to (i) interpret the nature of the innovation, (ii) legitimise adoption and (iii) mobilise the market to adopt the innovation, which together can increase adoption at all stages of diffusion [36]. Leveraging the organising visions lens, Kaganer et al. [16] developed the IT Legitimation Taxonomy for understanding the strategies leveraged by stakeholders to legitimise an innovation. This taxonomy enables the grouping of 26 micro-level legitimation strategies in accordance with the four types of legitimation they represent—cognitive, pragmatic, normative, and regulative. Consequently, researchers can explore legitimation building, the first stage of ILAM, within the broader lens of organising visions.

Blockchain is a nascent technology which has the potential radically change the IT, financial services and many other industries, and, as a result, generate substantial economic benefits. While previous research on technology adoption and innovation diffusion provides insights into the determinants of adoption for a number of IT innovations, this approach may not be suitable for blockchain for at least two reasons. Firstly, blockchain is a complex technology which transcends traditional organisational boundaries. By definition, blockchain is distributed and consequently individual adoption decisions are not sufficient. Instead, we need to understand the characteristics and interconnections between different organisations and adopters who themselves actively share their organising visions and key functionality for an innovation [37]. Secondly, blockchain is still a very nascent stage of development and there are still a number of unanswered questions regarding its legitimacy; thus, technology adoption is not an immediate consideration. Legitimacy is a essential step towards mobilisation, widespread adoption and the overall success of the technology [20]. As such, we argue that understanding the legitimation efforts of different business actors within the blockchain community, and indeed Bitcoin, is of greater importance. Consequently, this chapter does not seek to investigate the antecedents of organisational adoption of blockchain but rather to understand the social dynamics and strategic actions that organisations take to build legitimacy around an IT innovation as evidenced by their communications on Twitter.

RQ1. Which legitimation strategies do organisations invested in blockchain development adopt on Twitter?

Kaganer et al. [16] called for research using cross-sectional pattern analysis across different innovations. While Bitcoin is an innovation enabled by blockchain, we argue that actors in the Bitcoin community are fundamentally different than those in the blockchain community. Unlike the blockchain community, the use of Bitcoin as an anonymous and untraceable payments system for illegal transactions and money laundering has hindered the legitimacy of Bitcoin and threatened bans [38–43]. Furthermore, research suggests that social media plays a central role in building legitimacy in Bitcoin in a way that may not be the case for the wider blockchain market. For example, Mai et al. [44] suggest that social media is both important as a mechanism for information sharing on Bitcoin and a predictor of fluctuations in the Bitcoin price. Influential users may also trigger short-term changes in the Bitcoin market evaluation [44]. This influence, however, may, in fact, undermine Bitcoin adoption and legitimation efforts [45]. It is well-established that trust plays a key role in financial services [46, 47]. Bitcoin's role in dark marketplaces, such as the Silk Road, while bringing it to wider prominence, may have damaged its reputation and users' trust [48, 49]. As blockchain, independently of Bitcoin, is largely an enterprise technology with wider applications beyond a currency, legitimation strategies for blockchain should differ from those for Bitcoin. Consequently:

RQ2. Do legitimation strategies for blockchain differ from those adopted for Bitcoin on Twitter?

2.3 Research Design and Methodology

Our study focuses on how the media, IT, financial services and consulting firms use Twitter to build legitimacy around blockchain and Bitcoin. These organisations play an important role in this ecosystem for different reasons. The media is selected as it can affect perceptions of legitimacy in a positive or negative way [50, 51]. Blockchain is an IT innovation so IT firms have a clear role in the developing and advancing the technology. The financial services sector has heavily invested in blockchain already and such investments are expected to grow even more in the near future [52]. Finally, consulting firms typically provide market knowledge and bridge the gap between the IT sector and other sectors that may benefit for the technology [53, 54].

Our analysis focuses on the blockchain and bitcoin Twitter discourse over a 12-month period (1 July 2015 to 30 June 2016). Our dataset includes all English-

Table 2.1 Number of accounts and firms by dataset and actor type

Type	Blockchain dataset		Bitcoin dataset	
	No. of accounts	No. of firms	No. of accounts	No. of firms
Media	246	159	874	434
IT	96	66	130	103
Financial services	72	63	76	64
Consulting	51	24	27	16
Total	465	312	1107	617

language tweets containing the hashtags #blockchain or #bitcoin and/or the keywords "blockchain" or "bitcoin" and was compiled using the GNIP Enterprise API. GNIP data includes valuable information in the context of this study such as the content of the tweets, the type of message posted (i.e. original posts, retweets or replies), the timestamp, the screen name of user who posted the message, the Uniform Resource Locator (URL) of the user and its bio. We also retrieved Klout Score,[1] a measure of users' influence, using the dedicated API.

The initial dataset consisted of 11,956,529 tweets as generated by 368,213 discrete Twitter accounts. 9,291,748 of those tweets (77.7%) were original posts, 199,939 were replies (1.7%) and 2,664,781 (22.3%) were retweets. The list of users was then reduced to include only verified accounts with a Klout Score of 75 or higher (i.e. trustworthy and highly influential users). The remaining accounts were manually filtered to include only media, IT, financial services and consulting firms. For the remaining accounts, we only selected original tweets mentioning the hashtag #blockchain or the keyword "blockchain" for the blockchain dataset, or the hashtag #bitcoin or the keyword "bitcoin" for the Bitcoin dataset.[2] The final blockchain dataset comprised 3,653 tweets from 465 Twitter accounts representing 312 firms while the final Bitcoin dataset included 7207 tweets from 1107 Twitter accounts representing 617 firms. Table 2.1 provides an overview of the number of accounts and discrete firms by type and dataset.

A coding scheme was developed based on the IT Legitimation Taxonomy presented by [16] and adapted for a general IT context. More specifically, P1 and P2 were adapted for "quality of service" rather than "quality of medical care". Two independent coders interpreted the content of each post and classified each tweet into one of the 26 categories in the taxonomy or as "None" if no legitimation strategy was explicitly used. Inter-rater reliability with a Kappa co-efficient of 0.81, and

---

[1] Klout Score was a measure of the influence of a user across multiple social media platforms. Even though it was discontinued in May 2018, research suggests that it is a good proxy for credibility [55]. Klout score the ranges from zero to 100 and is based on three components across nine different social media platforms: (i) true reach, i.e. how many people a user influences; (ii) amplification, i.e. how much the user influences them; and (iii) network impact, i.e. the influence of the user's network [56, 57].

[2] All applied filters were not case sensitive.

a Cronbach's alpha of 0.90 was achieved for the blockchain dataset. For the Bitcoin dataset, the Kappa coefficient was equal to 0.78 while Cronbach's alpha was 0.88. The results of the manual classification were finally aggregated to explore the overall use of legitimation strategies and differences across different target actor.

2.4   Results

Figures 2.2 and 2.3 illustrate the percentage of tweets for each of the micro-level legitimation strategies for the blockchain and Bitcoin datasets, respectively. Some clear differences emerge. In the blockchain dataset, the four most common legitimation strategies are P13 (Alliance Field Level Actor), C7 (Diffusion Organisational), C3 (System Characteristics) and N2 (Normative—Transformational). More specifically, 22.42% of the tweets show clear attempts to build pragmatic legitimacy (P13), while another 26.36% of the tweets (13.85% with C7 and 12.51% with C3) are associated with cognitive legitimacy, and another 11.52% with normative legitimacy (N2). This suggests that the actors included in this study mostly aim to highlight the direct benefits associated with blockchain and to share general advocacy of blockchain-related initiatives from government or by established brands and alliances (pragmatic legitimacy). On the contrary, attempts to build cognitive legitimacy were mostly related to (i) positive market response to blockchain advancements, (ii) effective and potential real-life adoption or testing and (iii) how blockchain improves existing technological best practices. It is also worth noting that almost 15% of the tweets in our dataset are associated with some attempts to build normative legitimacy by pointing out the importance of adopting innovative technologies to adapt to potential change in the business environment [16].

In contrast, when one analyses the Bitcoin dataset, a significant percentage (36.56%) of the tweets are not associated with any particular legitimation strategy although a number of different ones are prominent. While C7 (Diffusion Organisational), P13 (Alliance Field Level Actor) and N2 (Normative—Transformational) were the most adopted strategies, together these account for only 25% of the tweets. Tweets with a clear focus on cognitive legitimacy mostly emphasise the increasing diffusion and ongoing development of Bitcoin (C7). These tweets are mostly related to (i) political endorsements, (ii) technology improvements, and (iii) the increasing number of Bitcoin transactions. 8.44% of Bitcoin tweets are associated with attempts to build pragmatic legitimacy by raising awareness of alliances between influential actors in the Bitcoin ecosystem (P13) such as financial services and technology companies, and regulators. However, on closer inspection, one can easily see that much of the alliance discourse is related to blockchain in a more general sense (e.g. *"IBM is becoming the biggest backer of a technology that underpins Bitcoin"*) rather than Bitcoin. Finally, 6.76% of the tweets in the Bitcoin dataset focus on the extent of changes Bitcoin may trigger (N2) and highlighting the potential disrupting effects of Bitcoin. Exemplar tweets for each type of legitimation and dataset are reported in Table 2.2.

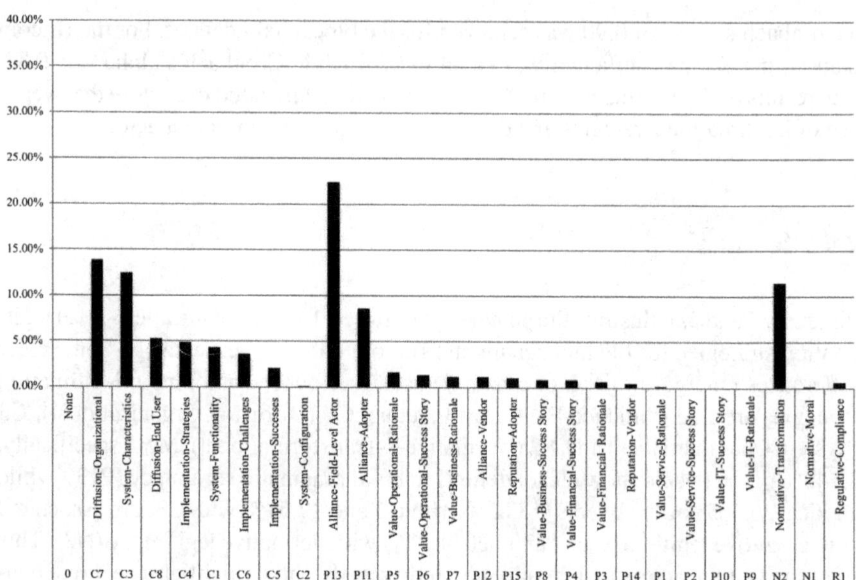

Fig. 2.2 Use of legitimation strategies—blockchain dataset

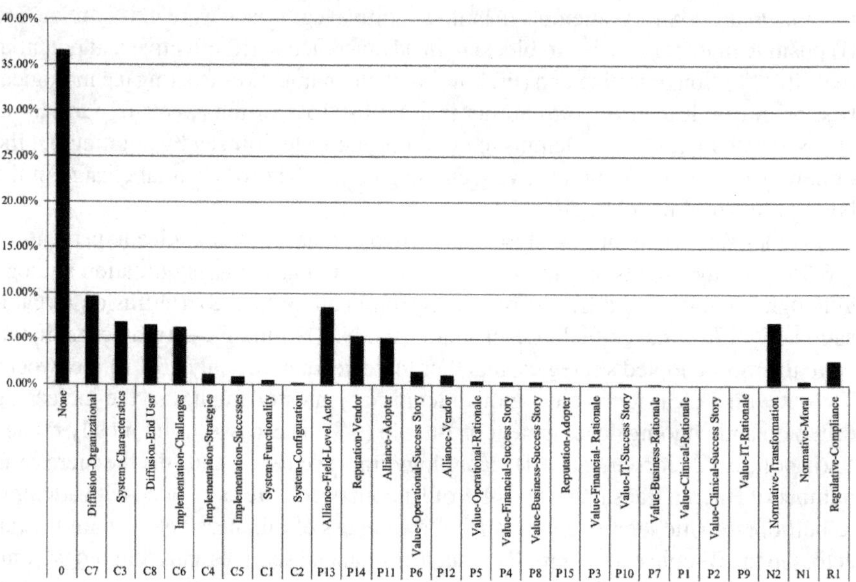

Fig. 2.3 Use of legitimation strategies—bitcoin dataset

Table 2.2 Examples of tweets by legitimacy type

Type	Dataset	Example
Pragmatic	Blockchain	Britain looking at blockchain for tracking taxpayer money
		This JPMorgan memo shows it is "aggressively" investing in blockchain tech and robotics
		Mizuho Financial Group partners with IBM to test #blockchain for settlements
	Bitcoin	Goldman Sachs put $50 million into bitcoin startup Circle and now it's coming to Europe
		Barclays Becomes First Big U.K. Bank To Accept Bitcoin
		Big names including Bain Capital and Mastercard are backing Barry Silbert's bitcoin venture
Cognitive	Blockchain	Westpac is the 43rd bank in a consortium developing the blockchain
		Blockchain spending expected to exceed $1 billion this year—report
		#Blockchain validates integrity and accuracy before accepting transactions. Once accepted transactions cannot be changed, reversed or erased
	Bitcoin	Banks could use bitcoin technology by next year
		Winklevoss twins' bitcoin firm sees trading surge
		How virtual currency #Bitcoin in gaining ground in #India
Normative	Blockchain	CEOs of large orgs say technologies like blockchain; #IoT will render their industries unrecognisable in 5 years
		Distributed ledgers like #Blockchain are likely to cause widespread disruption
	Bitcoin	Bitcoin to be major reserve currency by 2030: Research
		#Bitcoins: possibly the #future of money. Find out how to purchase them on @businessinsider
		A star Silicon Valley entrepreneur explains how bitcoin is going to change the world
Regulative	Blockchain	EU lawmakers to hold off from regulating blockchain for now
		"Regulators want #blockchain tech to succeed. They recognise that they need to evolve and understand". Steve Mollenkamp @IBM #Fintech Forum
	Bitcoin	No VAT on Bitcoin, rules ECJ, but capital gains still apply
		EU court rules that @Bitcoin exchanges don't have to pay taxes

Tables 2.3 and 2.4 report the number of each type of user and the corresponding level and types of activities. Most of the actors in the blockchain dataset are IT and media firms with 303 and 256 discrete users, respectively. Despite financial services and consulting firms being less represented in terms of the number of users when compared to IT and media firms, they are significantly more active, posting,

Table 2.3 Activity and micro-level legitimation strategy—blockchain dataset

Actor type	Descriptive		Strategies implemented	
			Code	No. of Tweets
Media	Tot. No. of tweets	1554	P13	406
	No. of users	246	C7	203
	Avg. User Activity	6.32	N2	155
			C3	144
IT	Tot. No. of tweets	636	P13	138
	No. of users	303	C3	135
	Avg. User Activity	2.10	C7	66
			N2	61
Financial Services	Tot. No. of tweets	853	P13	183
	No. of users	72	C7	135
	Avg. User Activity	11.85	N2	117
			C3	89
Consulting	Tot. No. of tweets	610	P13	102
	No. of users	51	C7	92
	Avg. User Activity	11.96	N2	89
			C3	88

on average, 11.96 and 11.85 tweets, respectively. This result may be interpreted as a further sign of the key role that financial services and consulting firms play in the blockchain ecosystem. In terms of micro-level legitimation strategies, all actors seem to focus on building pragmatic (P13), cognitive (C7 and C3) and normative legitimacy (N2) although the order and the relative frequencies vary slightly across different actors. IT companies, for example, tend to pay more attention to cognitive legitimacy than normative legitimacy.

In the Bitcoin dataset, media companies are the most represented with 874 unique users. This is not surprising considering the public interest and, consequently, the attention they pay to Bitcoin [9]. However, with 6.81 tweets per user, they are only the second most active group in the discourse. Financial services firms posted, on average, 10.04 tweets per user. Given the potential disruption to the payment sector, this is unsurprising. IT and consulting companies are the least active in the Bitcoin discourse with an average of 2.84 and 4.52 tweets per user, respectively. Taken together, these findings reflect the horizontal and wider application of blockchain as a technology platform, whereas Bitcoin can be considered a more vertical application. From a strategic perspective, the opportunities for blockchain are far greater than financial services, whereas for financial services firms, and banks in particu-

Table 2.4 Activity and micro-level legitimation strategy—bitcoin dataset

Actor type	Descriptive		Strategies implemented	
			Code	No. of Tweets
Media	Tot. No. of tweets	5953	None	2092
	No. of users	874	C7	556
	Avg. User Activity	6.81	P13	543
			N2	414
IT	Tot. No. of tweets	369	None	179
	No. of users	130	C7	33
	Avg. User Activity	2.84	C8	30
			N2	27
Financial Services	Tot. No. of tweets	763	None	339
	No. of users	76	C7	93
	Avg. User Activity	10.04	C3	64
			P13	48
Consulting	Tot. No. of tweets	122	C3	49
	No. of users	27	None	25
	Avg. User Activity	4.52	C7	10
			N2	9

lar, payment intermediation is one of the building blocks of the sector. The analysis of micro-level legitimation strategies reveals that only media and financial services companies appear to be attempting to build pragmatic legitimacy (P13). All actors seem mostly concerned with cognitive legitimacy, but with different preferences. While C7 is common to all actors, IT firms show greater interest in end user diffusion (C8), while financial services and consulting firms focus more on system characteristics (C3).

2.5 Discussion and Conclusion

This chapter answers calls for research which can offer insights on the acceptance and diffusion of blockchain and Bitcoin [58], and cross-sectional legitimation pattern analysis across different actors [16]. To do so, we leverage an inter-organisational perspective to extend IT innovation to a new focus on the broader blockchain and bitcoin community. The data set includes 617 firms across the four actor types, i.e. media, IT, financial services and consulting firms. Of the four actors examined, while most of them are linked to media and IT firms, financial service firms are more active in both discourses. Consulting firms were more active in the blockchain discourse than IT and media firms. We suggest this can be explained by the horizontal impact of blockchain in contrast to the vertical impact of Bitcoin, primarily on the financial services sector. This chapter, apart from adopting a broader focus than existing studies, accounts for the connected nature typical of emerging innovations such as blockchain and Bitcoin and recognises the key role played by a wide range of actors in building legitimacy around such innovations [37].

This study advances IS, blockchain and Bitcoin research further by applying the IT Legitimation Taxonomy to firm-level messaging. As discussed, this chapter answers to calls for studies on cross-sectional legitimation strategies analysis across different actors [16] and supports the extension of the IT Legitimation taxonomy to the blockchain, Bitcoin, and broader enterprise IS and financial innovation contexts. This chapter sheds light on to the types of legitimation strategies utilised by different actor types. In the blockchain discourse, there were two primary micro-level strategies utilised—pragmatic and cognitive legitimacy. This materialised through actors publicising the involvement of influential field-level actors (pragmatic legitimacy), describing positive market responses to blockchain and emphasising ongoing technology improvements (cognitive legitimacy), and describing the characteristics of blockchain that align with technological best practices (cognitive legitimacy). IT actors, in particular, are focused on system characteristics. In contrast, in the Bitcoin discourse, the largest portion of the tweets were not associated with any particular micro-level legitimation strategies suggesting that firms lack a clear communication strategy for Bitcoin, or possibly that it is seen as a limited and derivative discourse to the blockchain discourse. This potential lack of a clear communication strategy is further supported by a similar distribution of other tweets across other types of legitimation strategies, i.e. cognitive, pragmatic and normative. These results though are mostly driven by media companies. IT, financial services and consulting firms focus on building cognitive legitimacy by emphasising the growing adoption of Bitcoin and other system characteristics. This research reinforces the possibility to adopt the organising visions lens [19] and the IT Legitimation Taxonomy [16] for different technological innovations. The findings presented in this chapter clearly illustrate the value of leveraging the legitimation taxonomy to investigate and compare the micro-level legitimation strategies used by different actors to legitimise blockchain and Bitcoin on Twitter.

The findings presented in this chapter also extends the use of Twitter to understand legitimation efforts of established firms invested in the future of blockchain and Bitcoin. This represents a methodological advance for legitimation research which has a legacy of relying on single case studies. We argue that social media data represents a novel and reliable source of data for understanding efforts to build legitimation. Most of the existing studies utilise case study approaches (see, e.g., [16, 35]) but in so doing they limit their scope as legitimation efforts are typically examined retrospectively and the role of social processes and the role played by different actors is typically ignored. Our findings also supports the use of social media datasets for legitimation research as we demonstrate that all actor types actually engaged in legitimation strategies. Moreover, the findings support the use of the Integrated Legitimation Activity Model as a framework for investigating how innovation legitimacy is built within communities.

This research has practical implications for those seeking to legitimise blockchain, Bitcoin, and other IT innovations using Twitter. A relatively small number of micro-level strategies are used by firms and emphasise cognitive and pragmatic forms of legitimacy. For both entrepreneurs and their social media operators, the IT Legitimation Taxonomy presents a host of useful micro-strategies they may seek to leverage to legitimise blockchain and Bitcoin. The findings illustrate that the key actors actively engage in the blockchain and Bitcoin legitimation discourse to help interpret this complex innovation and mobilise the market to adopt. Actors in the blockchain and Bitcoin markets should consider the efficacy of using different legitimation strategies and how such strategies are operationalised in firm-level messaging on social media.

There are limitations to our research which illuminate areas ripe for further investigation. Our study is limited to one IT innovation, blockchain, and one derivative innovation, Bitcoin, one social network, one year, and four specific actors. Further research might include temporal legitimation analysis over several years (rather than just one year), compare enterprise actors and individuals, different innovations (e.g. other financial technologies) and networks, and using other qualitative and quantitative techniques such as, for example, network analytics, text mining or interviews. We hope these suggestions encourage computer science, information systems and finance researchers to further engage in multidisciplinary projects exploring the potential of blockchain, Bitcoin and legitimation.

References

1. Swan, M.: Blockchain: Blueprint for a New Economy. O'Reilly Media Inc. (2015)
2. Clohessy, T., Acton, T.: Investigating the influence of organizational factors on blockchain adoption. Industr. Manage. Data Syst. (2019). https://doi.org/10.1108/IMDS-08-2018-0365
3. Beck, R.: Beyond bitcoin: The rise of blockchain world. Computer **51**(2), 54–58 (2018)
4. Mougayar, W.: The Business Blockchain: Promise, Practice, and Application of the Next Internet Technology. Wiley (2016)
5. Nakamoto, S.: Bitcoin: A peer-to-peer electronic cash system. Bitcoin Project. https://bitcoin.org/bitcoin.pdf (2008)
6. Dodd, N.: The social life of Bitcoin theory. Cul. Soc. **35**(3), 35–56 (2018)
7. Barber, S., Boyen, X., Shi, E., Uzun, E.: Bitter to better–how to make bitcoin a better currency. In: International Conference on Financial Cryptography and Data Security. Springer, Berlin, Heidelberg (2012)
8. Lindman, J., Rossi, M., Tuunainen, V.K.: Opportunities and risks of Blockchain Technologies in payments—A research agenda. In: Proceedings of the 50th Hawaii International Conference on System Science (2017)
9. Glaser, F., Zimmermann, K., Haferkorn, M., Weber, M.C., Siering, M.: Bitcoin-asset or currency? revealing users' hidden intentions. Revealing Users' Hidden Intentions. In: Twenty Second European Conference on Information Systems (2014)
10. Zarifis, A., Efthymiou, L., Cheng, X., Demetriou, S.: Consumer trust in digital currency enabled transactions. In: International Conference on Business Information Systems, pp. 241–254. Springer, Cham (2014, May)

11. Cuen, L.: Bitcoin Usage Among Merchants Is Up, According to Data From Coinbase and Bit-Pay. Coindesk. https://www.coindesk.com/bitcoin-usage-among-merchants-is-up-according-to-data-from-coinbase-and-bitpay (2020)
12. Bank of International Settlements: BIS triennial central bank survey—Foreign exchange turnover in April 2019 (2019)
13. Saiedi, E., Broström, A., Ruiz, F.: Global drivers of cryptocurrency infrastructure adoption. Small Business Economics **1–54** (2020)
14. Stokes, R.: Virtual money laundering: The case of Bitcoin and the Linden dollar. Inform. Commun. Technol. Law **21**(3), 221–236 (2012)
15. Decker, C., Wattenhofer, R.: Bitcoin transaction malleability and MtGox. In: European Symposium on Research in Computer Security, pp. 313–326. Springer, Cham (2014)
16. Kaganer, E.A., Pawlowski, S.D., Wiley-Patton, S.: Building legitimacy for IT innovations: The case of computerized physician order entry systems. J. Assoc. Inform. Syst. **11**(1), (2010)
17. Hsu, P.F., Ray, S., Li-Hsieh, Y.Y.: Examining cloud computing adoption intention, pricing mechanism, and deployment model. Int. J. Inform. Manage. **34**(4), 474–488 (2014)
18. Lyytinen, K., Damsgaard, J.: What's wrong with the diffusion of innovation theory? In: Working Conference on Diffusing Software Product and Process Innovations. Springer, Boston, MA (2001)
19. Swanson, E.B., Ramiller, N.C.: Innovating mindfully with information technology. MIS Quarterly 553–583 (2004)
20. Wang, P., Swanson, E.B.: Launching professional services automation: Institutional entrepreneurship for information technology innovations. Inform. Organ. **17**(2), 59–88 (2007)
21. Rogers, E.M.: Diffusion of Innovations. Simon and Schuster (2010)
22. Davis, F.D.: Perceived usefulness, perceived ease of use, and user acceptance of information technology. MIS Quarterly 319–340 (1989)
23. Venkatesh, V., Morris, M.G., Davis, G.B., Davis, F.D.: User acceptance of information technology: Toward a unified view. MIS Quarterly **27**(3), 425–478 (2003)
24. Yusof, M.M., Kuljis, J., Papazafeiropoulou, A., Stergioulas, L.K.: An evaluation framework for Health Information Systems: human, organization and technology-fit factors (HOT-fit). Int. J. Med. Informatics **77**(6), 386–398 (2008)
25. Rodón, J., Sesé, F.: Analysing IOIS adoption through structural contradictions. Euro. J. Inform. Syst. **19**(6), 637–648 (2010)
26. Tornatzky, L., Fleischer, M.: The Process of Technology Innovation, p. 165. Lexington Books, Lexington, MA (1990)
27. Fichman, R.G.: The diffusion and assimilation of information technology innovations. In: Cincinnati, O.H. (ed.) Framing the Domains of IT Management: Projecting the Future Through the Past, pp. 1–42. Pinnaflex Educational Resources Inc. (2000)
28. Brown, S.A., Venkatesh, V., Goyal, S.: Expectation confirmation in information systems research: A test of six competing models. MIS Quarterly **38**(3), 729–756 (2014)
29. Prescott, M.B., Conger, S.A.: Information technology innovations: A classification by IT locus of impact and research approach. ACM SIGMIS Database **26**(2–3), 20–41 (1995)
30. Hirscheim, R., Klein, H.K.: Four paradigms of information systems development. Commun. ACM **32**(10), 1199–1216 (1989)
31. Suchman, M.C.: Managing legitimacy: Strategic and institutional approaches. Acad. Manage. Rev. **20**(3), 571–610 (1995)
32. Aldrich, H.E., Fiol, C.M.: Fools rush in–The institutional context of industry creation. Acad. Manage. Rev. **19**(4), 645–670 (1994)
33. Flynn, D., Hussain, Z.: Seeking legitimation for an information system: a preliminary process model. In: European Conference on Information Systems (2004)
34. Flynn, D., Du, Y.: A case study of the legitimation process undertaken to gain support for an information system in a Chinese university. Euro. J. Inform. Syst. **21**(3), 212–228 (2012)
35. Du, Y., Flynn, D.: Exploring the legitimation seeking Activities in an Information System Project (2010)

36. Swanson, E.B., Ramiller, N.C.: The organizing vision in information systems innovation. Organ. Sci. **8**(5), 458–474 (1997)
37. Brandt, C.J.: Toward a process view in adoption of interorganizational information systems. In: Proceedings of the 37th Information Systems Research Seminar in Scandinavia (IRIS 37) (2014)
38. Blundell-Wignall, A.: The Bitcoin question: Currency versus trust-less transfer technology. OECD Working Papers on Finance, Insurance and Private Pensions. Available via OECD. https://www.oecd.org/daf/fin/financial-markets/The-Bitcoin-Question-2014.pdf. Cited May 2017 (2014)
39. Bryans, D.: Bitcoin and money laundering: mining for an effective solution. Indiana Law J. **89**(1), 441–472 (2014)
40. European Central Bank (ECB): Virtual currency schemes. Available via ECB. https://www.ecb. europa.eu/pub/pdf/other/virtualcurrencyschemes201210en.pdf. Cited 10 March 2017 (2012)
41. Nee Lee, Y.: From China to Singapore, Asian countries are increasingly uneasy with the rise of bitcoin (2017). Available at: https://www.cnbc.com/2017/12/22/bitcoin-china-singaporejapan-issue-cryptocurrency-warnings.html. Accessed 12 January 2018; Nims, C. (2012). How universal one-click payments will change
42. Interpol: Digital currencies and money laundering focus of INTERPOL meeting. Available via Interpol. https://www.interpol.int/en/News-and-media/News/2017/N2017-002. Cited 11 May 2017 (2017)
43. Jung-a, S., Harris, B.: Bitcoin tumbles as South Korea plans trading ban. Financial Times. Available via Financial Times https://www.ft.com/content/0d5ff7d4-f67d-11e7-88f7-5465a6ce1a00. Cited 12 January 2018 (2018)
44. Mai, F., Bai, Q., Shan, Z., Wang, X.S., Chiang, R.H.: The Impacts of Social Media on Bitcoin Performance (2015)
45. Ciaian, P., Rajcaniova, M.: The digital agenda of virtual currencies: Can BitCoin become a global currency? Inform. Syst. e-Business Manage. **14**(4), 883–919 (2016)
46. Ennew, C., Kharouf, H., Sekhon, H.: Trust in UK financial services: A longitudinal analysis. J. Finan. Services Market. **16**(1), 65–75 (2011)
47. Hansen, T.: Understanding trust in financial services. The influence of financial healthiness, knowledge, and satisfaction. J. Service Res. **15**(3), 280–295 (2012)
48. European Banking Authority (EBA): EBA opinion on 'virtual currencies'. EBA/Op/2014/08, European Banking Authority (2014)
49. Yermack, D.: Is bitcoin a real currency? An economic appraisal. NBER Working Paper No. 19747, National Bureau of Economic Research. Available at: http://www.nber.org/papers/w19747. Accessed 12 Jan 2018 (2014)
50. Baum, J.A.C., Powell, W.W.: Cultivating an institutional ecology of organizations: Comment on Hannan, Carroll, Dundon, and Torres. Am. Sociol. Rev. **60**(4), 529–538 (1995)
51. Elsbach, K.D.: Managing organizational legitimacy in the California cattle industry: The construction and effectiveness of verbal accounts. Administrative Science Quarterly 57–88 (1994)
52. KPMG: Frontiers in Finance. For decision-makers in financial services. Issue #55. https://assets.kpmg/content/dam/kpmg/pdf/2016/05/frontiers-in-finance-may-2016.pdf (2016)
53. Hargadon, A.: How Breakthroughs Happen. The Surprising Truth About How Companies Innovate. Harvard Business School Press, Boston, MA (2003)
54. Sutton, R.: Weird ideas that spark innovation. Sloan Manage. Rev. **43**(2), 83–87 (2002)
55. Bode, L., Epstein, B.: Campaign Klout: Measuring online influence during the 2012 election. J. Inform. Technol. Politics **12**(2), 133–148 (2015)
56. Edwards, C., Spence, P.R., Gentile, C.J., Edwards, A., Edwards, A.: How much Klout do you have … A test of system generated cues on source credibility. Comput. Hum. Behav. **29**(5), A12–A16 (2013)
57. Rao, A., Spasojevic, N., Li, Z., Dsouza, T.:. Klout score: Measuring influence across multiple social networks. In: 2015 IEEE International Conference on Big Data (Big Data), pp. 2282–2289. IEEE, New York (2015, October)

58. Lindman, J., Tuunainen, V.K., Rossi, M.: Opportunities and risks of Blockchain Technologies-
    -A research agenda. In: Proceedings of the 50th Hawaii International Conference on System
    Sciences (2017)

Chapter 3
Using Blockchain in Intermittently Connected Network Environments

Souvik Basu, Soumyadip Chowdhury, and Sipra Das Bit

Abstract This chapter explores the possible integration of blockchain technology with intermittently connected networks, towards exploiting the utility and availability of blockchain technology in intermittently connected network environments. It also identifies the challenges of such integration and possible solutions using off-the-shelf technology. Finally, the chapter identifies open research areas in the domain of using blockchain in intermittently connected network environments that would foster new research avenues in both industry and academia.

3.1 Introduction

Blockchain [1–3] technology represents a technological innovation that is supposed to alter our lives in several aspects like the way we conduct business, manage assets, use machines, visit hospitals, cast votes, rent cars and even prove our identity. Apart from these traditional and urban applications, other specialized blockchain use cases can be disaster management, remote healthcare in developing countries, vehicular communications or even deep-space communications. However, in one hand, these specialized use cases are characterized by absence of traditional communication infrastructure, intermittent connectivity and disconnection of devices due to limitations of power, node mobility and sparse node density. On the other hand, the usage of blockchain is restricted by the user's access to end-to-end internet connection. This limitation restricts the use cases to access blockchain and prevents its adoption in intermittently connected network environments. In fact, reliance on the internet is

S. Basu (✉)
Heritage Institute of Technology, Kolkata, India
e-mail: souvik.basu@heritageit.edu

S. Chowdhury
University of Engineering and Management, Kolkata, India
e-mail: soumyadip.note@gmail.com

S. Das Bit
Indian Institute of Engineering Science and Technology, Shibpur, India
e-mail: sdasbit@yahoo.co.in

© The Author(s), under exclusive license to Springer Nature Singapore Pte Ltd. 2021
S. Patnaik et al. (eds.), *Blockchain Technology and Innovations in Business Processes*,
Smart Innovation, Systems and Technologies 219,
https://doi.org/10.1007/978-981-33-6470-7_3

becoming a critical concern for blockchain developers since network interruptions can cause serious harm to the blockchain ecosystem. This aspect has prompted the exploration of new avenues through which blockchain can be used even in intermittently connected network environments. To the best of our knowledge, this domain is very less investigated.

The blockchain technology uses crypto currencies such as Bitcoin and Ethereum [4, 5]. The digital nature of crypto currency requires online connectivity between two digital wallets to authorize transactions between the wallets. Theoretically, in absence of internet, a digital wallet could accept digital payment from another digital wallet offline, but that transaction would lack authorization. Solutions to this problem, proposed till date, include communicating with a node via SMS or satellite. However, SMS requires a fully functional network infrastructure, which cannot be expected in applications such as a post disaster scenario. Satellite is too expensive to be useful for the users without internet connections. Thus, it becomes imperative to look for other viable options for verifying blockchain transactions without requiring an end-to-end internet connection or traditional network infrastructure. Researchers in the field of networking suggest the usage of delay tolerant network (DTN) for establishing peer-to-peer communication networks using the WiFi Direct or Bluetooth interfaces of the smartphones, working in the DTN mode, in intermittently connected network scenarios [6–10]. Thus, the DTN protocols may suitably be enhanced to integrate them with blockchain so that verification of blockchain transactions may become feasible without requiring end-to-end internet connection or traditional network infrastructure.

This chapter proposes a mechanism that enables usage of blockchain technology in an intermittently connected network such as DTN. The mechanism develops an alternative way of broadcasting and authorizing blockchain transactions without having to rely on internet connectivity. It exploits the store-carry-forward feature of DTN nodes in conjunction with stationary relay nodes (e.g., smartphone, laptop), called DropBoxes to transmit and authorize transactions, hence enabling the usage of blockchain in DTN. The proposed mechanism is applied to the disaster management use case, based on the Ethereum platform using smart contracts in Solidity. The mechanism successfully integrates blockchain with DTN towards improving disaster management services in such an environment.

The remainder of this chapter is sorted out as follows. Section 3.2 gives a concise outline of DTNs and Sect. 3.3 records the simple components of blockchain system that are utilized in this work. Section 3.4 elaborates the mechanism proposed for integration of blockchain technology with DTNs using the disaster management use case. Section 3.5 explores open research areas in the domain of using blockchain in intermittently connected network environments. The chapter ends with a conclusion in Sect. 3.6.

3.2 Overview of Delay Tolerant Networks

Delay Tolerant Networks (DTNs) is proposed as an alternative mode of communication for areas where continuous end-to-end connectivity becomes unavailable [11, 12]. A DTN is a system of local systems which is demonstrated to be an overlay on top of the present existing systems including the web. DTN underpins interoperability among a lot of fluctuated systems by obliging entomb and intra organize interruptions and deferrals [13–15]. In spite of the fact that, DTNs were initially produced for interplanetary use, where delay-resistance is the best need, they have unmistakably more applications on earth [13, 16]. Figure 3.1 [12] shows a standard DTN that interfaces the Internet, different systems on earth and subnets over the nearby planetary group.

3.2.1 Features

Contrasted with customary Internet, mobile ad hoc networks and WLANs, DTNs have a one of a kind arrangement of highlights that are portrayed beneath [12]:

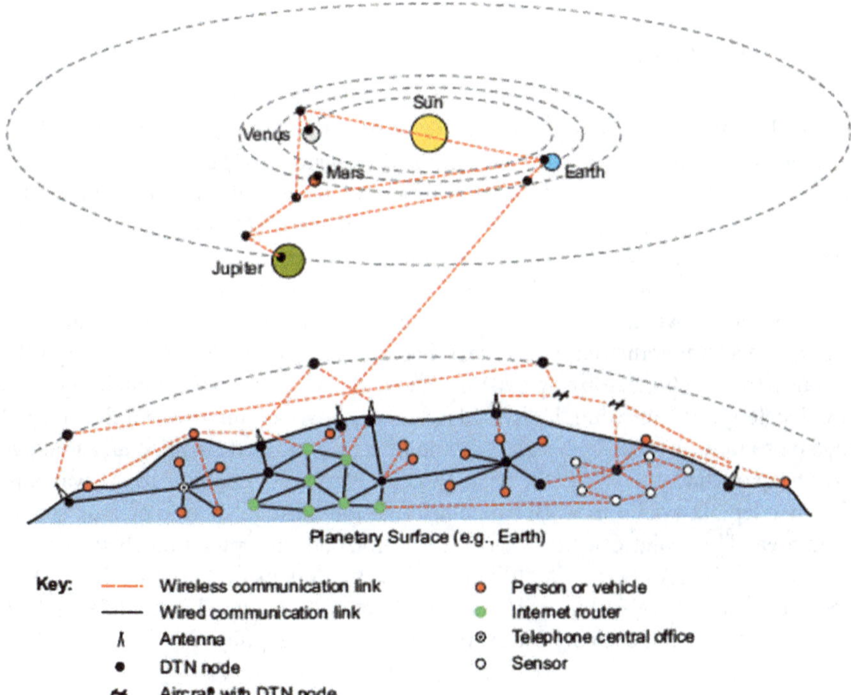

Fig. 3.1 A typical DTN [12]

Intermittent Connectivity: The nonappearance of a start to finish way among source and objective and the portability of hubs make successive system parceling in a DTN prompting irregular network.

High Delay: Notwithstanding discontinuous availability, long spread postponements among hubs and variable lining delays at hubs add to start to finish way delays.

Asymmetric Data Rates: Irregular network and very good quality to-end postpone prompts low and topsy-turvy up-down connection information rate.

Store-Carry-Forward: DTN follows a store-carry-forward mechanism for data transfer. In this mechanism, each node stores and carries data packets in the buffer, and forwards the duplicated data packets whenever the node meets at another node. Packets are carried till the expiry of the time-to-live (TTL).

Dynamic Topology: Ecological changes, vitality exhaustion or different disappointments, which brings about dropping out of the system, make the DTN geography very powerful.

Heterogeneous Interconnection: DTN is an overlay network organization that upholds interoperability among a large group of heterogeneous systems for transmission of non-concurrent message.

3.2.2 Architecture

As overlay system, DTN is planned to work over the current convention stacks in different system designs and gives a store-and-forward passage work between them when a hub truly present in at least two heterogeneous systems. Interoperability between these systems is cultivated by extraordinary DTN doors situated at their interconnection focuses. The DTN design incorporates the ideas of areas and DTN passages. Locale limits are utilized as interconnection focuses between heterogeneous system conventions and tending to families [11]. All the more officially, two hubs are in a similar district in the event that they can convey without utilizing DTN passages (by and large utilizing existing conventions nearby to the containing area). Few locale types (for example Web like, adhoc networks, intermittent disengaged, and so on) may advance and each example of a similar kind will execute a comparative heap of fundamental protocols [12]. DTN entryways compare to the waypoint idea that depicts a point through which information must go so as to pick up section to an area. This point can fill in as a reason for both interpretation (between area explicit encodings) just as a highlight implement strategy and control. In working over the vehicle layer, be that as it may, DTN doors are centered around dependable message steering rather than best-exertion bundle exchanging.

3.2.3 Routing

Routing is a key aspect in any kind of communication network including DTNs. DTN routing protocols [13, 16, 17] can be broadly classified as single copy protocols and multi copy protocols. In single copy protocols, messages are never replicated. Once a message has been delivered to an encountered node, it is deleted from the message buffer of the sender. Multi copy protocols can be comprehensively characterized into two sorts—limited and unlimited. In limited multi copy routing, a message is repeated for a fixed number of times and sent to various hubs in the system. In unlimited multi copy routing, there is no fixed breaking point on the quantity of replications of a given message that happen in the system. An outrageous case is flooding, where a given message is reproduced and sent to all the hubs in the system. Two popular single copy routing protocols are Direct-Delivery and First-Contact. Important multicopy routing protocols are Epidemic, PRoPHET, Spray-and-Wait, MaxProp, RAPID, SimBet, BubbleRap and Encounter Based Routing.

3.2.4 Applications

DTNs can be executed as a possible method of correspondence in various tested situations, for example, profound space systems. Some certifiable utilizations of DTN incorporate natural life following, post calamity correspondence systems, systems for far off zones or country regions in creating nations, vehicular systems and pocket-exchanged systems [16, 17]. These systems are dependent upon irregular network and detachment of hubs because of impediments of intensity, hub versatility, inadequate hub thickness, and technical disruptions.

Apart from these commonly found applications, DTNs can be applied to more demanding situations like disaster management and healthcare. On one hand, the systems administration research network has firmly proposed the utilization of DTNs for setting up crisis post debacle correspondence systems utilizing cell phones working in the DTN mode utilizing WiFi Direct or Bluetooth. Then again, there has been generous exploration on utilizing DTNs for giving option telemedicine framework zeroed in on performing distant emergency and prioritization of clinical consideration in far off districts.

3.3 Rudimentary Elements of Blockchain

Blockchain is a system in which a record of transactions made in Bitcoin or any other cryptocurrency are maintained across several computers that are linked in a peer-to-peer network. It is a distributed database at the same time. A blockchain is constantly growing as new sets of records/blocks are added to the chain. Thus, blockchain is a linked-list type data structure, which maintains the entire transaction history in terms of blocks [4, 5]. The overview of blockchain preliminaries relevant to our proposal is given below.

3.3.1 Blocks

In a blockchain, transactions are grouped into blocks so that they can be efficiently verified. Once a block is verified, they are added to the previous block. In a blockchain, each block is chained to its previous block through the use of a cryptographic hash. A change to a block forces a recalculation of all subsequent blocks, which requires enormous computation power. This makes the blockchain immutable, a key feature of crypto currencies like Bitcoin and Ethereum.

3.3.2 Transaction

A blockchain transaction can be defined as a small unit of task that is stored in a block. A typical transaction [4] from a sender to a recipient is shown in Fig. 3.2. A transaction has a unique identifier (TX) and consists of a set of inputs and outputs. Inputs to a transaction include previous transaction identifier (TX′) and signature of the user. Outputs from a transaction include the amount transferred to a recipient and the recipient's public key. To authorize spending the coins in the transaction input, the user should present its signature for the transaction and corresponding public key. Output of a transaction is used as input of the next transaction for verification. Transaction verification is distributed to P2P network nodes and only the valid transactions are recorded.

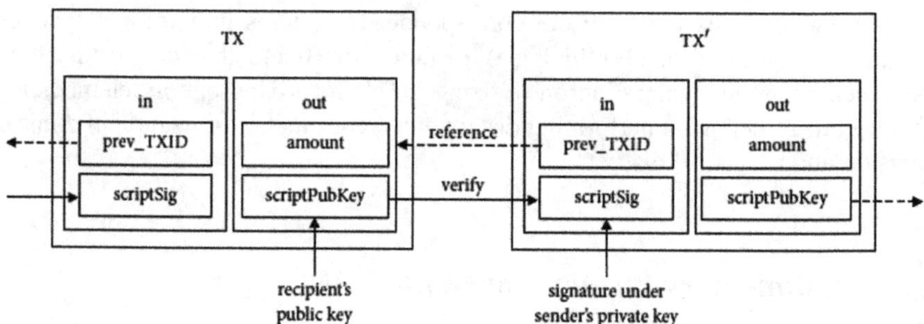

Fig. 3.2 Overview of blockchain transaction structure and spending [4]

3.3.3 Mining

Mining involves creating a hash of a block of transactions that cannot be easily forged, protecting the integrity of the entire blockchain without the need for a central system. Some computers storing the entire blockchain are called miners. When a transaction is made over a blockchain the transaction information is thus put on a block and gets recorded. All the miners crack a cryptographic riddle so as to locate the correct cryptographic hash for the hash. When an miner makes sure about the hash, it is added to the blockchain and must be confirmed by different nodes on the system in a cycle known as consensus. At the point when a miner effectively confirms and makes sure about the hash, it is given a reward [5].

3.3.4 Ethereum and Smart Contracts

Ethereum [5] is an open source, public, blockchain-based distributed computing platform. Ether is the cryptocurrency generated by the Ethereum platform as a reward to mining nodes for computations performed and is the only currency accepted in the payment of transaction fees. Ethereum provides a decentralized virtual machine, the Ethereum Virtual Machine (EVM), which can execute scripts using an international network of public nodes. "Gas" is the fee, or pricing value, required to successfully conduct a transaction or execute a contract on the Ethereum blockchain platform.

A smart contract is a piece of code that is running on Ethereum to execute credible transactions without any third parties. It is a self-executing contract between network participants, without the need of traditional legal contracts. The contracts are high-level applications that are compiled down to EVM bytecode and deployed to the Ethereum blockchain for execution. An Ethereum user creates a contract, and pushes the data to that contract so that it can execute the desired command.

3.4 Integration of Blockchain Technology with DTNs

In this section, we elaborate our proposed mechanism for integration of blockchain technology with DTNs for a disaster management use case.

When a disaster strikes, collection of the victims' needs at different relief shelters and its dissemination towards different stake holders (e.g. relief agencies, Govt. agencies, NGOs and disaster managers) are the crucial activities for minimization of injury, loss of life and property smash up [7–9]. Such information can help different disaster management stakeholders to appropriately coordinate and manage relief activities. Since enormous pieces of the communication facilities get totally debilitated as a result of the fiasco, it gets hard to gather and communicate the necessities from various distant and blocked off havens to design the salvage activity appropriately. The networking researchers have emphatically proposed the utilization of DTNs for setting up a post disaster communication system utilizing the WiFi Direct or Bluetooth interfaces of the cell phones, conveyed by volunteers and salvage vehicles, working in the DTN mode [6–10]. Such networks can be used to collect and transmit post disaster shelter needs in a peer-to-peer basis through the volunteers carrying smartphones. Now, to make the shelter needs available to all stakeholders it becomes imperative to put up these needs as unchallengeable and globally accessible records, for efficient disaster management. Several emergency response organizations have proposed putting up the shelter needs on a blockchain to create a shared and immutable system of records. This calls for a necessary integration of blockchain technologies with DTNs for uploading shelter needs collected over a DTN to a blockchain network. The major challenges of such integration can be summarized as below:

- Locally validating the shelter needs collected from volunteers in absence of internet
- Converting the DTN blocks (containing validated shelter needs) to blockchain compatible blocks
- Uploading blocks to the blockchain network

With these challenges in mind, we propose a mechanism for integrating the blockchain technology with DTNs. With the help of this integration, the post disaster shelter needs can be collected from remote and far flung shelters, validated, converted to blocks and finally uploaded to a blockchain network. As a result, the needs become immutable and globally accessible to all disaster management stakeholders, which in turn, leads to efficient disaster relief. The entire mechanism is schematically shown in Fig. 3.3.

The following sub-sections elaborate the system components and the steps of the proposed integration mechanism.

3.4.1 System Components

During and after disasters, victims normally take shelter in nearby safe areas. A control station is setup at nearby resource-rich areas [6–10]. Our proposed mechanism maps the above scenario where the smartphones carried by the volunteers and supply vehicles communicate by forming a post disaster communication network characterized by DTN. The network used in our proposed mechanism comprises of five major components, whose functions are described below.

Shelter—A shelter consists of a shelter-node (e.g., a laptop with Wi-Fi Direct). Emergency manager(s) present in a shelter uploads shelter needs to the shelter-node which, in turn, broadcasts this information towards the volunteers that are working in and around it.

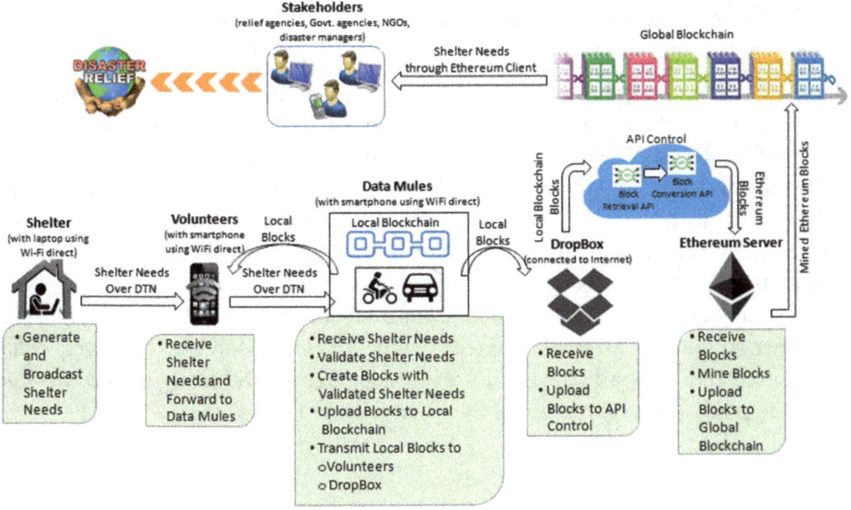

Fig. 3.3 Schematic diagram of the proposed mechanism

Volunteer—Volunteers carrying smartphones with Wi-Fi Direct, are the forwarders who move across the disaster stricken area, collect shelter needs from the shelter-nodes, exchange these needs among themselves using the DTN routing protocol (e.g. PRoPHET [13]) and eventually transmit them to the data mules.

Data Mule—Supply vehicles equipped with computing devices having Wi-Fi Direct, are data mules that move across the disaster stricken area at a speed greater than that of the volunteers, collect shelter needs from the volunteers, process them and deliver the processed needs to the nearest DropBox using the DTN routing protocol PRoPHET [13].

DropBox—Stationary relay nodes (e.g., smartphone, laptop having LTE/5G internet connectivity, possessing sufficient storage and computation capacity) that receive and store processed shelter needs from the data mules passing by it and forward them to the API control forward further processing. DropBoxes are deployed at high priority locations, based on the uDBdep scheme [6] to maximize data transfer. Internet connectivity is essential at the DropBoxes for transmitting the processed shelter needs to the API control running on a cloud server.

Ethereum Server—Ethereum server runs the open source, public, blockchain-based distributed computing system. It mines blocks and upload them to the global blockchain.

3.4.2 Generation and Transmission of Shelter Needs

The emergency managers, who coordinate relief activities in a shelter, collect the shelter needs like the amount of food, medicine, etc. and upload them to the shelter-node therein. The shelter-node stores these values which are then broadcast towards the volunteers carrying smartphones. The volunteers store, carry and forward the shelter needs to the data mules roaming around the disaster affected area, either directly using Wi-Fi Direct or in multiple hops through other volunteers using DTN routing. In our proposed mechanism, the volunteers assure the data securing as well. Every volunteer puts a digital signature on every message block before passing it to another volunteer, ensuring volunteer's authentication. Moreover, from this chain of signatures the forwarding route for every message can be easily tracked back to detect the point of discontinuity for every undelivered message.

3.4.3 Need Validation and Creation of Local Blockchain

Data mules receive the shelter needs from volunteers and validate them for possible errors using block mining. Post validation, the shelter needs are converted to blocks which are uploaded to a local blockchain. These blocks are also forwarded to the volunteers and the nearest DropBox using WiFi Direct. This enables the volunteers to have a coherent view of the shelter needs at a particular point in time. The DropBox receives the blocks from the data mules and uploads it to an API control running on a cloud server.

3.4.4 Block Conversion Through API

We develop an Application Program Interface (API) and an Application Binary Interface (ABI) that play the pivotal role in deploying the blocks containing shelter needs, in the form of smart contracts on the Ethereum network. Our specific contributions are:

- Developing an API for establishing the connection between the DTN leveraged post disaster communication network and Ethereum server
- Developing an ABI for converting the DTN blocks (containing validated shelter needs) to Ethereum blocks using Solidity contracts
- Deploying the converted blocks on global blockchain using JSON-RPC protocol

In this work we use programming languages like Java and Solidity; frameworks like Spring Boot, Spring Rest, Spring MVC, Spring Data and Spring Security; and technologies like Maven-Building tool, Metamask-Crypto wallet and gateway to blockchain wallet and Rinkeby-Ethereum Test Network. Initially, a Maven & Spring

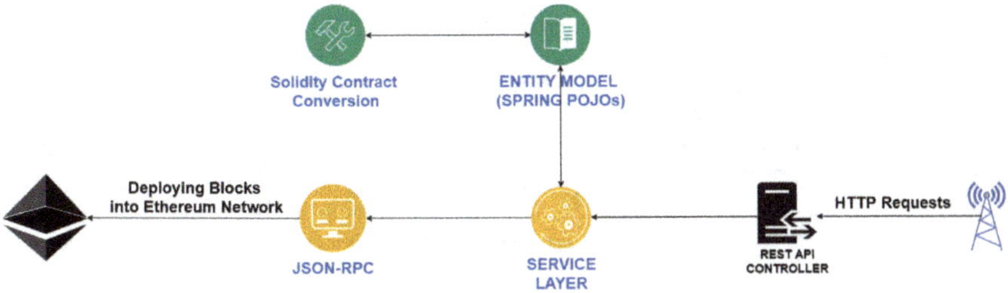

Fig. 3.4 The designed API control

Boot project is created which adds all dependencies (Spring Rest, MVC, Spring Data, Spring Security, JSON-RPC, ABI, Lombok, Web3J, Slf4j, etc.) and creates the pom.xml & application.properties files. After creating the project successfully, we develop the API that works as the bridge between the local blockchain data and Ethereum network using Rest API Controller. The Rest API Controller creates a link ("/accept-local blocks-from-dropbox") with support of Get Request and Post Request. The DropBox can send data to the API control using this API.

After receiving the blocks of the local blockchain from the DropBox, the API control frames the blocks as Pojos (Plain old Java Objects). All Pojos are then converted into Solidity contracts using the developed ABI. These Solidity contracts are framed as Ethereum blocks. Thus, blocks of the local blockchain are converted into Ethereum blocks. These blocks are deployed to the Ethereum server in real time, using JSON-RPC protocol. Figure 3.4 elucidates the functioning of the developed API control.

3.4.5 *Mining Blocks and Deploying to Global Blockchain*

After the blocks are deployed on the Ethereum server, the server mines all blocks using its own miners, and pays a certain amount of gas to the miners as a payment for mining the blocks. After every successful mining, all DTN blocks are deployed into the Global Ethereum Network, which is completely immutable. All stakeholders would be able to access the blocks consisting of shelter needs using Ethereum clients into their local machine, which can be further used to fittingly facilitate, oversee and channelize their assets.

The sequence diagram in Fig. 3.5 depicts the sequence of actions in the proposed integration mechanism.

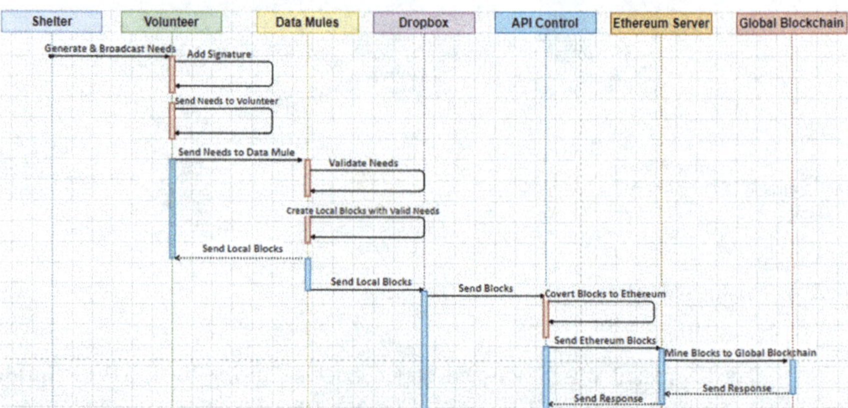

Fig. 3.5 Sequence diagram for the proposed integration mechanism

3.5 Open Research Areas

In this section we give directions towards future research on successful exploitation
of our proposed integration/access of blockchain with DTNs in different application
domains.

Incentive Scheme for DTNs—The network performance of node-cooperation based
networks like DTNs deteriorate by selfish nodes that do not participate in message
forwarding due to their concerns on resource consumption. Therefore, participating
nodes should be awarded satisfactory incentives to compensate its resource depletion
for cooperation. Although, the available incentive schemes look like a plausible solution to this problem, most of them either rely on central trusted authorities or do not
use an explicit digital currency that is provably secure. Blockchain, a decentralized
secure computerized record of unchanging exchanges, is an appealing methodology
for tending to the motivating force difficulties in shared systems that need focal
confided in specialists. Bitcoin, the Blockchain based digital money, make it conceiv-
able to devise functional credit based motivation plans for such systems. Although
authors in [9] propose a scheme for such incentivization, no actual implementation
is suggested. In such works, our proposed integration mechanism can be utilized
to devise blockchain based incentive scheme that may use Bitcoin for incentivizing
nodes in DTNs for cooperation.

Post Disaster Resource Management—One of the major concerns in any emergency
relief operation is appropriate allocation of scarce emergency relief materials to the
affected community. To address the issues of allocation transparency, resource diversion, etc., it becomes imperative to have a document for providing an unchallengeable
and globally accessible record of relief requirement vis-à-vis allocation for efficient
relief management. Emergency response organizations like UNICEF, FEMA have
recommended the adoption of blockchain technology to create such a shared and

unchallengeable document of records [18, 19]. However, the usage of blockchain is restricted by the availability of end-to-end internet connection which, as mentioned earlier, may not be available in a post-disaster scenario. Thus, our proposed integration mechanism may be used to propose a blockchain leveraged post disaster relief allocation system over DTN that works in challenged network environments. Such a system, if designed, will be able to execute smart contracts for validating relief requirements to mitigate resource diversion, forecasting the exact demand and enumerating precise utilities of relief items to alleviate incorrect allocation.

Applications in Rural Areas of Developing Countries—Rural areas of developing countries like India are still largely devoid of Internet connectivity. As a result, providing internet based services like remote healthcare and e-Governance to the rural communities become challenging. Our proposed integration/access of blockchain with DTNs can be utilized in providing these types of services to the rural population.

Remote Patient Monitoring—Remote patient monitoring (RPM) is a developing area and is transforming the healthcare sector [20]. RPM, using wearable devices, makes use of IoT technology to transfer patient data to health services providers. As a result, assurance on secured transmission of exact information to health administrations suppliers at the ideal time is a significant achievement factor for RPM. Also, health information is very delicate and should consistently be stayed up with the latest and secure. Adulterated or lost health records and solutions can cause significant issues when exact data is required. Blockchain can help in guaranteeing the exactness of information recorded by RPM gadgets and encouraging secure and opportune exchange of patient information. As the access to private blockchains is restricted, such sensitive data gets adequate protection. A system on a private blockchain may be designed to execute smart contracts for analyzing patient health data. Upon analyzing the data, smart contracts can issue alerts if the data values are beyond certain thresholds. The framework would then be able to mention that well-being objective fact accessible to pertinent partners through the open blockchain. Along these lines, the framework can store understanding data in a solid brought together framework where their information is secure and can't be altered. However, in absence of end-to-end internet facilities in rural areas, such systems cannot be implemented. Therefore, our integration mechanism can be applied to collect patient health data from the RPM devices, used by rural patients, through a DTN based IoT, get them analyzed by a local private blockchain. Results of such analyses can be transferred, once again over DTN, to a nearby DropBox with internet facility which in turn can make them available to relevant stakeholders via the public blockchain.

Rural e-Governance—Blockchain innovation has exhibited its possibility and importance in online business. Its utilization is currently being reached out to new regions, identified with electronic administration [21]. Blockchain innovation is being presented in numerous nations, for an assortment of e-Governance administrations, similar to enrollment of portable and unfaltering resources, protected innovation, wills, social insurance, medical care information, and benefits frameworks.

Tried blockchain arrangements are accessible to direct sell-offs, to advance straightforwardness of the public and nearby financial plans, to make sure about solid vote including in races, to make crowdfunding stages empowering speculators to follow consumptions on their ventures. However, lack of end-to-end internet connectivity rural regions hinders the implementation of blockchain for e-Governance. In such regions, our proposed integration mechanism can be applied to collect user data over a DTN environment and perform a first level validation using local blockchains without using internet. This validated information can be uploaded to the public blockchain once internet is available to create global and immutable records.

3.6 Conclusion

This chapter proposes a mechanism to integrate blockchain technology with DTNs to exploit the immutability and accessibility of blockchain in an intermittently connected network. It develops an alternative way of broadcasting and authorizing blockchain transactions without having to rely on internet connectivity. It uses the store-carry-forward feature of DTN nodes in conjunction with stationary relay nodes (e.g., smartphone, laptop), called DropBoxes, to transmit and authorize transactions, hence enabling the usage of blockchain in intermittently connected network environments. The proposed mechanism is applied for the disaster management use case, based on the Ethereum platform using smart contracts in Solidity. It successfully integrates blockchain with DTN environment towards improving disaster management services in absence of end-to-end to the internet. The chapter also gives an insight into the possible future research on successful utilization of our proposed integration of blockchain with DTNs in different application domains.

References

1. Nakamoto, S.: Bitcoin, a peer-to-peer electronic cash system. https://bitcoin.org/bitcoin.pdf (2008)
2. Antonopoulous, A.M.: Mastering Bitcoin—Programming the Open Blockchain. O'Reilly Media (2017)
3. Sengupta, J., Ruj, S.R., Das, B.S.: A comprehensive survey on attacks, security issues and blockchain solutions for IoT and IIoT. J. Netw. Comput. Appl. **149** (2020)
4. Park, Y., Sur, C., Rhee, K.-H.: A secure incentive scheme for vehicular delay tolerant networks using cryptocurrency. Secur. Commun. Netw. 1–13 (2018)
5. Understanding Blockchain: A Beginners Guide to Ethereum Smart Contract Programming, https://www.codemag.com/Article/1805061/Understanding-Blockchain-A-Beginners-Guide-to-Ethereum-Smart-Contract-Programming. Accessed 9July 2020
6. Das, N., Basu, S., Das, B.S.: Efficient dropbox deployment towards improving post disaster information exchange in a smart city. ACM Trans. Spat. Algorithms Syst. **6**(2), 1–18 (2020)
7. Basu, S., Roy, S.: Secured categorization and group encounter based dissemination of post disaster situational data using peer-to-peer delay tolerant network. Adv. Intell. Syst. Comput. **328**, 747–755 (2015)

8. Chakrabarti, C., Roy, S., Basu, S.: Intention aware misbehavior detection for post-disaster opportunistic communication over peer-to-peer DTN. Peer-to-Peer Netw. Appl. **12**, 705–723 (2019)
9. Chakrabarti, C., Basu, S.: Blockchain based incentive scheme for post disaster opportunistic communication over DTN. In: Proceedings of International Conference on Distributed Computing and Networking 2019, ACM Digital Library, pp. 385–388 (2019)
10. Basu, S., Roy, S.: A global reputation estimation and analysis technique for detection of malicious nodes in a post-disaster communication environment. In: Proceedings of International Conference on Applications and Innovations in Mobile Computing 2014, IEEE Xplore, pp. 179–185 (2014)
11. Roy, A., Acharya, T., DasBit, S.: Quality of service in delay tolerant networks: a survey. Comput. Netw. **130**, 121–133 (2018)
12. Fall, K., Farell, S.: DTN: an architectural retrospective. IEEE J. Sel. Areas Commun. **26**(5), 828–836 (2008)
13. Roy, A., Bose, S., Acharya, T., DasBit, S.: Social-based energy-aware multicasting in delay tolerant networks. J. Netw. Comput. Appl. **87**, 169–184 (2017)
14. Fall, K.: A delay-tolerant network architecture for challenged internets. Proc. SIGCOMM **2003**, 27–34 (2003)
15. Sun, W., Liu, C., Wang, D.: On delay-tolerant networking and its application. In: International Proceedings of Computer Science and Information Technology, vol. 51, pp. 238–244 (2012)
16. Basu, S., Roy, S., Das, B.S.: Introduction. In: Reliable Post Disaster Services over Smartphone Based DTN. Springer Smart Innovation, Systems and Technologies, vol. 137 (2019)
17. Massri, K.: Routing protocols for delay tolerant networks: a reference architecture and a thorough quantitative evaluation. J. Sens. Actuator Netw. **5**(2), 1–28 (2016)
18. https://www.ledgerinsights.com/fema-blockchain-disaster-insurance-parametric/
19. https://www.unicef.org/innovation/blockchain
20. Hathaliya, J., Sharma, P., Tanwar, S., Gupta, R.: Blockchain-based remote patient monitoring in healthcare 4.0. In: International Conference on Advanced Computing 2019, IEEE Xplore, pp. 87–91 (2019)
21. Terzi, S., et al.: Blockchain 3.0 Smart Contracts in E-Government 3.0 Applications. https://arxiv.org/ftp/arxiv/papers/1910/1910.06092.pdf (2019)

Chapter 4
Slaying the Crypto Dragons: Towards a *CryptoSure* Trust Model for Crypto-economics

Blockchain Versus Trust: The Expert's View of the Crypto Scammers

Stephen Castell

Abstract In order to move towards a *CryptoSure* regime, systems based on blockchain and distributed ledger technology have to give up the fiction of 'consensus trustless mechanisms', and, if they are to provide socially and legally acceptable security, privacy and trust, within the *Rule of Law*, have to accept the necessity of a *Trusted Third Party*. For example, the General Data Protection Regulation (GDPR) includes in its provisions Article 17, the *Right to be Forgotten*, which could potentially be a formidable barrier to the ubiquitous introduction of cryptographic blockchain software and technology outside of a *CryptoSure* regime. The investment mania that there has been for blockchain technology, with much money having gone into bitcoin and other cryptocurrencies, blockchain, smart contracts and distributed ledger technology, has been significantly fuelled by the 'black cash' of drug-dealers, money-launders, traffickers and the like. *Crypto Dragons*, the many and varied *Financial Disputes over Crypto Assets*, have arrived, with legal actions mounted by those defrauded increasing. This is not surprising, given that in Q2 2019 alone, misappropriation of cryptocurrency funds netted criminals some \$4.26 billion. The foundations of global digital currencies go back well before the Satoshi bitcoin paper of 2008, and those early digital e-commerce visions did not require a cryptographic blockchain 'mining', or 'distributed consensus', existential model and were not intentioned of being so readily riven with the criminal black market profiteering of money-launders, scammers and fraudsters that bedevil much current cryptocurrency activity. Looking ahead, Facebook's *Libra* digital currency could establish a new global e-commerce paradigm much closer to the pre-bitcoin electronic cash visions, and one more compliant with the existing norms and customs of the *Rule of Law*, where a responsible *Trusted Third Party*, in this case, Facebook, is fundamental. Cryptocurrencies apart, some blockchain applications more generally are likely here to stay, and the majority will be robust implementations by established major corporations, with most of us, as consumers, hardly needing to know any of

S. Castell (✉)
CASTELL Consulting, Witham, UK
e-mail: stephen@castellconsulting.com

     49
S. Patnaik et al. (eds.), *Blockchain Technology and Innovations in Business Processes*,
Smart Innovation, Systems and Technologies 219,
https://doi.org/10.1007/978-981-33-6470-7_4

the details. For the properly cautious ICT expert and professional, when considering the use of blockchain for any proposed use case, the 'fundamental things apply'. There is always the need for Trusted Third Parties, and for probative Electronic Evidence. A key point in any court trial will be *examination of the Digital Evidence* and, although a Crypto Asset may essentially be 'decentralised digital vapour', a Court of Law can make a binding order to get forensic traction on it, because of the legally well-established *Obligation of Disclosure*. This article concludes with a *Checklist* giving practical, generally applicable wording for an effective *Digital Asset Disclosure* exercise.

4.1 Introduction: Blockchain and the Right to Be Forgotten

> Blockchain technology introduces permanence and immutability into the digital world. … the technological revolution that commoditizes trust … Trust normally has to be enforced via laws, courts, … fallible institutions. Replacing these with disinterested cryptography promises a revolution in the way we enable trust. … [This brings up] the right to be forgotten. A law that grants individuals, under some circumstances, the right to demand of websites that they remove information about themselves. However, in a distributed consensus system like blockchain, enforcing the right to be forgotten becomes technically impossible. …

Júlio Santos, November 6th, 2017.[1]

The *Right to be Forgotten* could potentially be a formidable barrier to the ubiquitous introduction of computer and communications systems applications based on cryptographic blockchain software and technology. The General Data Protection Regulation (GDPR), in force from May 25, 2018, includes in its provisions Article 17:

> http://www.privacy-regulation.eu/en/article-17-right-to-erasure-'right-to-be-forgotten'-GDPR.htm.

> "Right to erasure ('right to be forgotten')"… (e) the personal data have to be erased for compliance with a legal obligation in Union or Member State law to which the controller is subject; …

With the 'permanence and immutability' of data records written to the blockchain being emphasised as one of its fundamental, key features, in a wide range of use cases where acquisition, processing and recording of *personal data* is critical blockchain could possibly be structurally unable to be compliant with Article 17, Right to Erasure, of GDPR. The *Commission nationale de l'informatique et des libertés* (CNIL), the independent French administrative regulatory body whose mission is to ensure that data privacy law is applied to the collection, storage, and use of personal data, has identified this fundamental issue:

---

[1] 'Forever on the Chain' https://hackernoon.com/forever-on-the-chain-c755838dfc79.
   Júlio Santos, November 6th, 2017. https://lifeonmars.pt.

> … one of the characteristics of blockchains is that the data registered on a blockchain cannot be technically altered or deleted: once a block in which a transaction is recorded has been accepted by the majority of the participants, that transaction can no longer be altered in practice. … technical solutions … should be examined by stakeholders in order to solve this issue. The CNIL … questions their ability to ensure a full compliance with the GDPR…
>
> As a reminder, a blockchain can contain two categories of personal data:
>
> The identifiers of participants and miners:
>
> Each participant has an identifier comprised of a series of alphanumeric characters which look random, and which constitute the public key to the participant's account. This public key is linked to a private key, known only by the participant…
>
> The CNIL therefore considers that this data cannot be further minimised and that their retention periods are, by essence, in line with the blockchain's duration of existence.
>
> Additional data (or payload):
>
> Besides the participants' identifiers, the additional data stored on the blockchain can contain personal data, which can potentially relate to individuals other than participants and miners.
>
> As a reminder, the principle of data protection by design (Art 25 of GDPR) requires the data controller to choose the format with the least impact on individuals' rights and freedoms.

Others have proposed potential technical solutions, for example:

> The Workaround … Storing personal data on a blockchain is not an option anymore according to GDPR. A popular option to get around this problem is a very simple one: You store the personal data off-chain and store the reference to this data, along with a hash of this data and other metadata (like claims and permissions about this data), on the blockchain.
>
> Andries Van Humbeeck, November 21, 2017.[2]

There is also a technician's view that, in regard to interpreting and implementing 'erasure' in practice, simply 'putting data beyond use' electronically will satisfy the standards for GDPR data privacy. This would mean that, for example, setting record 'delete' flags, 'losing' cryptographic keys, or overwriting hash tables, will be sufficient to qualify as 'erasure'.

However, I consider this too weak to satisfy what is intended and stipulated by Article 17 GDPR. If Article 17 had sought to provide only for 'putting data beyond use' it would have said so. The people doing the drafting would have been aware of, amongst other things, the established legal precedents and court orders on:

- Data records, recording media and destruction (and proof/certification thereof);
- Corporate, industry and professional standards as regards record retention and destruction; and
- Statutes providing requirements and guidelines for public bodies as regards citizens' records disposal.[3]

---

[2] 'Solutions for a responsible use of the blockchain in the context of personal data' https://www.cnil.fr/sites/default/files/atoms/files/blockchain.pdf.
'The Blockchain-GDPR Paradox' https://medium.com/wearetheledger/the-blockchain-gdpr-paradox-fc51e663d047.

Andries Van Humbeeck, November 21, 2017. https://theledger.be/.

[3] https://bitsonblocks.net/2016/02/29/a-gentle-introduction-to-immutability-of-blockchains/.

The word chosen in Article 17 of GDPR is 'erasure', and its intention and meaning is something clear, stringent and strong. If GDPR had intended 'erasure' just to mean, or include, 'putting data beyond use', or even 'deletion', in the usual technical sense that these terms are used and implemented in electronics and computer data technology practice, it would have made that, too, clear. GDPR was years in the drafting, with many highly-qualified legal and technical people involved, globally, in intensive discussions and reviews, before finalisation. 'Erasure' and 'erased', being the actual words carefully enacted in the GDPR, have many clear synonyms in English: *'Erasing': eradicating, obliterating, destroying, abolishing, removing, shredding, disposing of, wiping out, dissolving, doing away with, getting rid of...*

From an expert point of view, where digital data recorded on servers, or electronically held, copied, distributed and communicated in computer and communications media, systems and networks are concerned, 'erasing' can even mean, for true efficacy in practice, 'returning to a free molecular state' by way, for example, of 'burning, consuming in flames'.

It follows that anyone implementing applications or systems using a blockchain, given the foundational, inherent 'permanence and immutability' of its data records, where such records may contain personally identifiable details of a 'data subject', will do so at risk of not being physically or verifiably able to comply with Article 17 GDPR, and thus potentially subject to the significant financial and other penalties available and arising thereunder.

It may be considered that there will be little likelihood of requests, whether to companies or organisations holding or processing systems and databases containing personally identifiable details of 'data subjects', or to the courts, for applicant data subjects to be 'forgotten'. A few years back the possibility of widespread use of such requests may have seemed fanciful, but since the *Cambridge Analytica* allegations—that this data analytics firm used personal information harvested from more than fifty million Facebook profiles, *without the data subjects' permission*, to build a system that could target US voters with personalised political advertisements based on their psychological profile—anyone using social media, for example, is now well aware of the right not to have personal data used for purposes for which they were not originally, and freely, provided.

Furthermore, even before the coming into force of GDPR the English Courts had upheld such a critical request:

---

https://www.coindesk.com/blockchain-immutability-myth/.
https://www.forbes.com/sites/yec/2017/05/04/debunking-blockchain-myths-and-how-they-will-impact-the-future-of-business/#583e1d815609.
https://www.records.nsw.gov.au/recordkeeping/advice/retention-and-disposal/destruction-of-records.
https://ct.wolterskluwer.com/resource-center/articles/three-simple-rules-record-retention.
https://www.proshred.com/documentation-retention-policies-potential-ramifications/.
http://www.pbi.org/docs/default-source/default-document-library/10778_minimizing-commercial-litigation-risks-(2019).pdf?sfvrsn=0.
'Minimizing Commercial Litigation Risks ... 9. Implement a Document Retention Policy...'.
https://www.scality.com/blog/fuhgettaboutit-the-gdpr-right-to-erasure/.

https://www.theguardian.com/technology/2018/apr/13/google-loses-right-to-be-forgotten-case

Google loses landmark 'right to be forgotten' case Jamie Grierson Ben Quinn 13 Apr 2018

Businessman wins legal action to force removal of search results about past conviction

A businessman has won his legal action to remove search results about a criminal conviction in a landmark "right to be forgotten" case that could have wide-ranging repercussions. … the claimant … was convicted more than 10 years ago of conspiracy …[4]

4.2 The New 'Crypto-economy'—A Fraudsters' Playground?

Despite that the GDPR Article 17 risk to systems implemented using a blockchain, in use cases where personal data is to be recorded, presents a potentially serious implementation difficulty, there has been an investment mania for crypto-algorithmic blockchain technology, with far more money having gone into—gambled on—bitcoin and other cryptocurrencies, blockchain, smart contracts and distributed ledger technology than even into artificial intelligence (AI). It has in the past seemed that almost every other Millennial was involved with an initial coin offering (ICO) or initial token offering (ITO). With just a 'white paper', little or no investment due diligence, and taking advantage of a regulatory vacuum, this 'crypto tribe' raised billions in real legal tender, 'fiat currencies'.

This substantial finance-raising has been used to fund fantasy coins and tokens, with no obvious economic utility or asset value, in the hope of developing and successfully launching a plethora of brave new business and social ideas, products and services, heralded by enthusiasts as a whole new 'crypto-economy'. A few of these may prove to be commercially-successful, reputable, significantly disruptive game-changers, and usher in the possibility of a new global 'crypto-economy' paradigm.

---

[4]https://internetofbusiness.com/solid-pods-how-sir-tim-berners-lee-aims-to-inrupt-the-web-from-within/.

The position, security, ownership, handling, (mis)use, etc. of 'personal data' and who should profit from such data, is becoming central to the future social media and digital economy, globally. Sir Tim Berners-Lee has introduced a new breed of '…*personal online data stores, or Pods, that contain the wealth of information people generate, and are their exclusive property* …', as part of the future 'Web 3.0'.

My own www.Zykme.net, a new P2P cross-platform comms App (hybrid, beta test version), designed to provide instant private one-to-one secure transfer of personal data using a unique proprietary *one-time Zykword code* protocol, is consistent with this Pod Philosophy. The *Zykme* App does not demand the user's email address, nor any other 'logon' ID data; and, as made clear at the foot of its '*Your ZykPod History of Contacts Received*" page, '*NOTE Unlike other social media, this unique Zykme App, which securely provides instant two-way peer-to-peer communication, does not and will not acquire, store, process, analyse, use nor pass on to any third party any of your entered data. Using Zykme, your personal 'My details' information as 'Sender' remains completely under your control, and, at entirely your own decision and choice, is privately and confidentially shared and exchanged between you and your selected '…Receiver' alone, when you press 'Share info*".

But so far, it has often been discovered that ICOs/ITOs, cryptocurrency 'mining', and crypto-coin trading exchanges have tended to have been significantly fuelled or taken over by the 'black cash' of drug-dealers, money-launders, traffickers and the like, and in a substantive not-easily-reversible way.

The 'Q2 2019 Cryptocurrency Anti-Money Laundering Report' from *Ciphertrace* revealed that misappropriation of funds 'from cryptocurrency users and exchanges netted criminals and fraudsters approximately $4.26 billion in aggregate'. This is in the context of the amount of cryptocurrency traded in September 2019 on crypto-trading exchanges being over $500 billion (down from nearly $800 billion in June 2019), with Hong Kong-based exchange *Binance* reporting that, in the last two years, it alone made over $1 billion of profit.

Many of those yearnings for the putative 'crypto-economy', for example, Millennials let down after the post-2008 credit crunch by governments, the banks and educational system, have tended to disregard any need to be subject to Know Your Client (KYC) and Anti-Money Laundering (AML) strictures, and may not have been too worried from whence came their ICO money, how it was actually going to be (accountably) spent, or whether it could even possibly result in a viable business.

It is worth being reminded that the foundations of global digital currencies go back well before the Satoshi bitcoin paper of 2008. The early pioneering international digital economy e-commerce visions did not require a cryptographic blockchain 'mining', or 'distributed consensus', existential model. And they certainly were not intentioned by any thought or expectation of becoming so readily riven with the criminal black market profiteering of money-launders, scammers and fraudsters that apparently increasingly bedevil much—but of course not all—current cryptocurrency activity.

David Chaum, in a scientific paper of 1983, is reckoned to be the first to describe digital money. His proposal used cryptography to create a blind, digital signature to make money anonymous, and he founded a company in 1989 that invented the virtual currency *DigiCash*. But it had a hard time commercially, with a 1999 article in *Forbes* summing it up as: 'A beautiful idea for a beautiful new world with one problem: nobody wants it. Not the banks, not the dealers and above all, not the customers. E-commerce is flourishing, but as it turns out, the customer's Mastercard and Visa are his preferred currencies'.

Milton Friedman, the economist, said in 1999: 'One thing we are still lacking and will soon develop is reliable e-cash—a method by which money can be transferred from A to B on the Internet without A knowing B and vice versa'. Even earlier than these, I myself put forward, twenty-five years ago, a new, *disintermediated* wholly digital cash currency, as set out in my letter published in July 1995 in *Computing* magazine:

> … As cybertrading grows, the new, powerful common electronic trading currency will be 'owned' by no single physical nation state, central bank institution, economic or political grouping. … the Electronic Cash Unit.

And, long before Millennials were even born, my fictional article, '*Ye Nom De Das Geld*', in the December 1971 issue of *GONG* (the student magazine of the University of Nottingham), went even further with my vision of a 'Post-Purse Paradise':

> Brother and sisters, I welcome you to the post-purse paradise. … Geld is in heaven, all's well with the world. … Cromstock and I first mooted the possibility of an Economic Reformation taking place in Britain in The Journal Of Comparative Economics during … 1969. … to put into practice … the tenets of the Quasicurrency Theory which I had been formulating over the preceding twenty-five years…

Looking ahead, Facebook has plans for its *Libra* digital currency that so worry regulators they are seriously considering trying to prevent it happening; but it may already be too late, and *Libra* could be 'unstoppable'. Facebook has some 2.4 bn monthly users across its various apps with users on at least one of these apps every day. Furthermore, *Libra* may turn out to have little to do with any putative 'crypto-economy' and could establish a powerful new digitalised global e-commerce paradigm much closer to the pre-bitcoin electronic cash visions of early digital currency thought-leaders and entrepreneurs—and a paradigm more comfortably compliant with the existing human society and regulatory norms and customs of the *Rule of Law*, where a responsible *Trusted Third Party*, in this case, Facebook, is fundamental, and pivotal.

For all these reasons many of the current species of cryptocurrencies—not excluding bitcoin—may in their present manifestations fade away, and/or the hoped-for 'crypto-economy' may at some point even be regulated out of existence.[5]

---

[5] '*What the ECU stands for*', Stephen Castell, Letter in *Computing*, 20 July 1995.

'*Ye Nom De Das Geld*', Stephen Castell, *GONG* Magazine, December 1971, pp. 16–18.

https://www.arachnys.com/2019/10/22/addressing-the-aml-risks-of-cryptocurrencies/.

'Addressing the AML risks of cryptocurrencies OCTOBER 22, 2019 BLOG With the recent explosion in cryptocurrencies, from the early beginnings of bitcoin back in 2009 through to J.P. Morgan testing their own digital coins for institutional clients in 2019, there still remains serious unanswered questions about the money laundering risks they bring to banks, consumers and regulators. Ciphertrace's 'Q2 2019 Cryptocurrency Anti-Money Laundering Report' makes some stark revelations. It claims that theft, scams and other forms of misappropriation of funds 'from cryptocurrency users and exchanges netted criminals and fraudsters' approximately $4.26 billion in aggregate. … Dr. Stephen Castell, an independent FinTech consultant, admits that there are few innocent investor protections to fall back on: 'This is essentially the case worldwide today, and it looks like it will continue that way for the foreseeable future'. He reminds us that there is a need to keep everything in perspective, suggesting that "the actual, and potential, total global 'crypto' business for banks and other financial institutions is tiny—in the less than 1% area". So, with the increased anti-money 1 laundering (AML) risks associated with blockchains and cryptocurrencies, he believes it's right for the compliance departments of banks to proceed cautiously, if at all…'.

'…Traditional exchanges around the world will be looking at Binance's latest quarterly results with envy, as in the last two years it has made over $1billion of profit. CEO of Binance, Changpeng Zhao, … established Binance only in 2017. Binance raised $15 million via an Initial Coin Offering (ICO) and CZ is reported to be worth $1.2 billion. Binance, based in Hong Kong, is different from its competitors which, apart from Huobi (Singapore), are based in the USA, e.g. Coinbase (San Fran), Kraken (San Fran), Bittex (Las Vagas) and Bitbox (NYC). The amount of Cryptos that were traded in September 2019 on exchanges like Binance was still over $500 billion-down from nearly

4.3 Blockchain: Sceptical ICT Professionalism and Legal Due Diligence

Cryptocurrencies apart, however, some blockchain applications more generally are likely here to stay. The majority of these will be serious, robust implementations, by established major corporations, with most of us, as consumers, hardly needing to know about the technical, legal or operational details. It seems clear that, within a few years, a widespread settled, but vigorous and continually innovating, 'blockchain applications industry' may be in place, one perhaps bearing little resemblance to the frantic cryptocurrency 'bandit territory' landscape of today.

For the properly-cautious ICT expert and professional, when considering the use of blockchain for any proposed use case, the 'fundamental things apply'. This caution is an essential part of being a skilled professional applying knowledge and experience to assess the most appropriate tools and technologies for a given (business or other) application's requirements. The savvy ICT expert bears in mind, for example, that not only are there no finalised international/ISO standards for blockchain (the eight standards in development under ISO/TC 307, due out in 2020, have not yet been completed), but also there is far more to specifying, designing, developing, testing, deploying and maintaining an appropriate complete QA-assured system than just 'the blockchain component'.

And whether to use blockchain as a component *at all* for a given business/system requirement is a critical feasibility exercise that the seasoned professional will know is vital. Any duly diligent ICT systems engineer may therefore conclude, on an experienced expert assessment, that many things can be achieved just as effectively by other means. He or she will carefully and responsibly consider all the pros and cons to ensure that the non-expert customer/client/investor/employer (to whom a professional fiduciary duty is owed) gets the most suitable, 'fit for purpose', secure, robust and performant system available. Ideally, this will also take properly risk-assessed competitive advantage of any—and not just crypto, or blockchain—new developments in technologies, tools, methodologies and processes, always consistent with the budget/price willing to be paid, of course.[6]

$800 billion in June 2019. According to the website Coin.Market, there are now over 260 different crypto exchanges …'.

Digital Bytes, Weekending 26th October 2019, TeamBlockchain Ltd.

http://www.teamblockchain.net/.

https://hackernoon.com/the-amazing-story-of-cryptocurrencies-before-bitcoin-fe1b0e55155b.

'The Amazing Story of Cryptocurrencies Before Bitcoin Marcell Nimfuehr, 14 October 2019 What—you exclaim with disbelief. Cryptocurrencies before bitcoin? Yes, indeed. Do not get me wrong, Bitcoin was the first blockchain-based currency. But by far not the first purely digital money. That one has a colourful history of dreams, prosecution and failure…'.

https://www.dcforecasts.com/libra-coin-news/chinese-crypto-czar-facebooks-libra-might-be-unstoppable/.

'Chinese Crypto Czar: Facebook's Libra 'Might Be Unstoppable' September 20, 2019 Stefan'.

[6]*Blockchain Standards*

https://www.iso.org/committee/6266604.html.

Furthermore, the legal status of blockchain cryptocurrency, smart contract and distributed ledger technology is not clear, or uncontentious. In the USA, there is already ICO litigation on foot.[7] Having been involved in advising on ICOs, I have encountered some significant tensions and challenges between the crypto-enthusiastic, blockchain technical specialist, and the sober business development objectives of, and the professional due diligence to be done for, the putative ICO-issuing company owner or managing executive.

Consider, for example, this scenario: a proficient, high-profile, software engineering entrepreneur and thought-leader, let us call him Joshua, a US citizen, a highly experienced and imaginative technical and regulatory expert working in the blockchain and cryptocurrencies field, is developing and launching various Initial Coin Offering ventures and services. Joshua asserts 'nobody knows more about how to do this work in the right way, in compliance with every single rule and regulation, than I do'. There is a substantial going-concern OTC-listed company, let us call it XYX-CAP, Inc. ('XYX-C'), which is poised to do an ICO, designed, led, promoted, launched and actioned-to-market by Joshua.

ISO/TC 307 Blockchain and distributed ledger technologies.

Scope: Standardisation of blockchain technologies and distributed ledger technologies.

8 ISO standards under development under the direct responsibility of ISO/TC 307.

34 Participating members 12 Observing members.

'Blockchain—The Legal Implications of Distributed Systems', The Law Society HORIZON SCANNING August 2017, 12 pages.

Blockchain Patents

https://worldwide.espacenet.com/searchResults?ST=singleline&locale=en_EP&submitted=true&DB=&query=blockchain.

https://www.cnbc.com/2019/03/25/bank-of-america-skeptical-on-blockchain-despite-having-most-patents.html.

https://www.americanbar.org/groups/intellectual_property_law/publications/landslide/2017-18/march-april/patentability-blockchain-technology-future-innovation/.

https://thenextweb.com/hardfork/2019/03/13/data-china-is-patenting-all-the-blockchain-tech-despite-banning-cryptocurrency/.

The many blockchain patents—though perhaps not yet all granted, let alone challenged—may illustrate a difficulty that the ISO Working Parties could encounter in trying to define 'International Standards', which are essentially meant to be 'Open Source'.

https://www.infosys.com/Oracle/white-papers/Documents/integrating-blockchain-erp.pdf.

http://www.primechaintech.com/assets/docs/PT-BSC-0_4.pdf.

'Primechain Technologies Blockchain Security Controls Version 0.4 dated 21 October, 2017'

https://www.dlapiper.com/en/uk/insights/publications/2017/06/blockchain-background-challenges-legal-issues/.

'2 February 2018 Blockchain: background, challenges and legal issues By: John McKinlay Duncan Pithouse John McGonagle Jessica Sanders (née Turner)'

https://www.forbes.com/sites/laurashin/2016/05/10/looking-to-integrate-blockchain-into-your-business-heres-how/#4986f47f1a15.

'10 May 2016 Looking To Integrate Blockchain Into Your Business? Here's How Laura Shin Companies … are sprinting to begin adopting blockchain… But many are doing so simply because of fear of missing out, without a clear understanding of how it can be useful …'.

[7] https://www.prnewswire.com/news-releases/silver-miller-files-class-action-lawsuit-against-monkey-capital-and-its-principal-daniel-harrison-for-alleged-fraudulently-promoted-and-aborted-initial-coin-offering-300574019.html.

The following queries and issues arise:

1. If the XYX-C coin created by this ICO is likely to be deemed by any relevant (US or other) regulatory or law-enforcement authority to be 'asset-backed', and equivalent to *issuing a security*, would it not be advisable to *seek securities regulatory approval for this ICO before it is publicly launched*? If so, what exactly is the relevant and correct 'securities regulatory approval' to be sought, with whom, where, etc. and how does one go about that, correctly, accurately and timeously?

2. Joshua says *'It's very important to be aware that this is an open community blockchain project. This necessarily involves launching something that will have the XYX-C name attached to it in perpetuity, but giving up exclusive control of what it becomes'*. If the CEO of XYX-C is not wholly comfortable with this, are there any sensible steps that XYX-C can take to protect its name, brand and trademark to counter (or at least ameliorate) 'giving up control of what it becomes'? If so, what, and how, and at what cost to put it in place?

3. Suppose this ICO goes badly wrong at some point, and either the XYX-C company, or the public at large investing in the XYX-C coin, claim they have lost money, or otherwise been damaged by taking part in its launch, and also claim that Joshua made misrepresentations, and was negligent/fraudulent, and thus seek reparations or, worse, criminal prosecution, what can he do to avoid, or protect against, that possibility, or its consequences, *at the outset*, i.e. *before* the ICO is launched publicly? Are there any sensible legal and practical protective steps he can take?[8]

4.4 The Need for Trusted Third Parties and for Probative Electronic Evidence

Commissioned by the UK's *Central Computer and Telecommunications Agency* (CCTA, H M Treasury), and funded by the five major UK Departments of State (Defence, Home Office, Treasury, Foreign Office and Industry), I carried out in

---

'… CORAL SPRINGS, Fla., Dec. 20, 2017 … www.SilverMillerLaw.com … actions currently Pending against the Coinbase, Kraken, and Cryptsy exchanges … Monkey Capital fraudulently promoted an ICO that violated numerous state and federal securities laws…'.

https://www.silvermillerlaw.com/david-silver/2017/12/20/silver-miller-files-class-action-law suit-monkey-capital-principal-daniel-harrison-fraudulently-promoted-aborted-initial-coin-off ering/.

'…As ICOs have become more frequently used as a fundraising tool for start-up blockchain technology companies, so too has fraud upon cryptocurrency investors become more frequent; and Monkey Capital appears to have been a prime example of the harm investors can suffer …See the Class Action Complaint: Hodges, et al. v. Monkey Capital LLC, et al. …'.

https://www.silvermillerlaw.com/wp-content/uploads/2017/12/2017-12-19-DE-1-CLASS-ACTION-COMPLAINT.pdf.

[8] 'CASTELL—Legal Due Diligence for Initial Coin Offering 07Feb2018.pdf'. Available privately from the author, on application.

1990 a major study <u>on the admissibility of computer evidence in court and the legal reliability/security of IT systems</u>. This was done in the main by field research (with civil servants, lawyers, techies, 'records management' personnel and business folk), high-level interviews, confidential brainstorming seminars and what would today be called 'focus groups'. The work resulted in my *Verdict Report*, with a follow up *Appeal Report*, both UK government confidential (and classified).

That major study was later permitted by the UK government to be published, in edited/sanitised form, as *The APPEAL Report* (1990), still seen by many expert professionals and practitioners as definitive in the critical field of the legal standing and trustworthiness of computer software and systems, and digital or electronic evidence derived from such systems. It concluded with what became known as:

> Castell's (First) Dictum: "You cannot secure an ontologically unreliable technology by use of an ontologically unreliable technology".

It is vital for any operational computer system, and, not least, one purporting to provide goods, services, currencies, communications, etc. to the public, upon which the public relies, to have one or more *Trusted Third Party* (TTP) standing behind it and responsible for it, given the ontological unreliability of computer technology, and the associated need for disclosure of probative *Electronic Evidence* and computer 'documents' when (not if!) disputes arise. Electronic Evidence has become widely acknowledged to be based on the concept of a transactional *chain of trust*, and I also identified in 1993, in wide-ranging research for the Commission of the European Community, the latter's dependency on *Trusted Third Party Services* (TTPS):

> A Trusted Third Party is an impartial organization delivering business confidence, through commercial and technical security features, to an electronic transaction. It supplies technically and legally reliable means of carrying out, facilitating, producing independent evidence about and/or arbitrating on an electronic transaction. Its services are provided and underwritten by technical, legal, financial and/or structural means.

Thus. TTPS are provided and underwritten not only by technical, but also by legal, financial and structural means and are operationally connected through chains of trust (usually called certificate paths) in order to provide a *web of trust*: the whole legal, financial, technical and organisational structure being what we might identify as a robust working example of, and call simply, an *Implementation of the Rule of Law*.[9]

---

[9] *The APPEAL Report*, Dr Stephen Castell, 1990, May, Eclipse Publications, (ISBN 1-870771-03-6).

'Code of practice and management guidelines for trusted third party services', S. Castell, INFOSEC Project Report S2101/02, 1993.

'Green paper on the security of information systems', Commission of the European Community, ver. 4.2.1, 1994.

See also in:

'Security Issues On Cloud Computing', Pratibha Tripathi, Mohammad Suaib; Department of Computer Science and Engineering, Integral University, Lucknow, Uttar Pradesh, India.

International Journal of Engineering Technology, Management and Applied Sciences http://www.ijetmas.com/ November 2014, Volume 2 Issue 6, ISSN 2349-44761. Available from:

4.5   Towards a *CryptoSure* Trust Model
for Crypto-economics

Bearing in mind, on the one hand, this established foundational background as regards the essential legal, financial, technical and organisational structures required, by way of TTPS, to implement robustly and assuredly the *Rule of Law*, and, on the other, the equally fundamental truth of the ontological unreliability, with current machine architectures, of computer software and systems, it is clear that systems based on blockchain and distributed ledger technology have to give up the fiction of 'consensus trustless mechanisms', and, if they are to provide socially and legally acceptable security, privacy and trust, within the *Rule of Law*, have to accept the necessity of a *Trusted Third Party*.

Regrettably, the evidence appears to be that most of those in what may be called the 'crypto-evangelical wing' of the blockchain community are wilfully, and certainly in some cases, fraudulently, blind to this truth. As was identified and elaborated earlier in this article, thefts, scams, frauds and other forms of misappropriation of funds in the cryptocurrency and blockchain sector persist, thrive and grow. It is patently clear that the supposed trust and reliability of 'blockchain only' systems and services said to be provided, technically, cryptographically, by virtue solely of the 'distributed consensus' algorithm, much-vaunted and promoted by the 'crypto-evangelists', is frankly a dangerous nonsense.

As I argue in a forthcoming article *I, Bitcoin: My Merkle Tree, My Vacuity* (in finalisation, to be submitted to a leading financial journal):

> … I, Bitcoin have no utilitarian rationale. The stark reality is that I exist only *to* exist. … I owe my creation to no *Invisible Hand*. … and you will have discerned the insight of *the wisdom of the absolutely essential ingredient for freedom—a faith, and trust, in free markets, and in the Trusted Third Party that the Rule of Law provides in the fair and equitable underpinning of those free markets.* Freedom is impossible without this faith, and this trust; and such confidence is therefore impossible without the *Trusted Third Party*, and the *Rule of Law*. I, Bitcoin, therefore reflect with all the more anguish on my own arid emptiness. Do come forward, if you can, with an idea as to how the *Invisible Hand*, and thus market freedom and faith, underpinned by *Trusted Third Parties*, and secured by the *Rule of Law*, can be made manifest in the utilitarian, purposive functioning of me, Bitcoin! Without emergence

---

https://www.researchgate.net/publication/272945014_Security_Issues_On_Cloud_Computing.

The *Draft Convention on Electronic Evidence* has recently been published, in the Volume 13: 2016 issue of the *Digital Evidence and Electronic Signature Law Review*. It is authored by Stephen Mason (http://www.stephenmason.eu/), a barrister of the Middle Temple and a recognised authority on electronic signatures and digital evidence, with contributions by Dr Stephen Castell. To obtain and review the *Draft Convention on Electronic Evidence*:

1.   Go to http://journals.sas.ac.uk/deeslr/issue/view/336/showToc.
2.   See 'Documents Supplement' at foot of contents; click on 'Draft Convention on Electronic Evidence' to see Abstract: http://dx.doi.org/10.14296/deeslr.v13i0.2321.
3.   Then, click on 'PDF' (http://journals.sas.ac.uk/deeslr/article/view/2321/2245) to download the full text of the draft convention.

of this strongly-founded 'free, responsible and trusted market' model for a practically useful crypto-economic existence, *I will remain in existential vacuity*. And I fear it follows that I, Bitcoin, will inevitably wither, become valueless, and die.

Pending emergence, if ever, of such a strongly-founded, TTPS and *Rule of Law*, 'free, responsible and trusted market' model for practically useful crypto-economic activities and operations, and given that those in the 'crypto-evangelical' community (whether that is mainly a community of energetic, visionary, financially imaginative, disruptively creative, honestly aspiring entrepreneurs, or, simply scammers, thieves, fraudsters and liars, it matters not!) persist in promoting a false trust in the *solus* 'distributed consensus' algorithm, I have envisioned a practical, pragmatic *Trust Model for Crypto-economics* with components which I tentatively propose could in the interim provide what I have christened a **CryptoSure** regime.

Components of the *CryptoSure* regime

1. Creation of the *CryptoSure Foundation*, with charitable status, run and overseen by a Board of Trustees drawn from senior reputable expert practitioners in, for example, the legal, management, ICT, business, financial, banking and regulation fields of professional expertise, particularly (but not exclusively), with experience in cryptocurrency, crypto assets, blockchain, DLT, Smart Contract, etc. endeavours, companies and ventures.

2. The *CryptoSure Foundation* will grant operating membership to any individual, business, company, organisation who applies, subject to a qualification procedure which will check and validate that the putative member demonstrably has:

   - Appropriate insurance cover in place (e.g. professional indemnity and product liability).
   - Met a quality and cybersecurity assurance audit of systems design, development, build, code, technology, etc.
   - Mechanisms and protocols in place to comply with KYC and AML standards and regulations.
   - Registered with relevant regulatory authorities (e.g. financial and/or securities regulators), where applicable.
   - Deposited an audited, signed and dated business plan in the *CryptoSure Repository* (which must include, e.g. CV details of directors, promoters, technical specialists etc.).
   - Provided satisfactory business, bank, lawyer, accountant and personal references, on the venture, and on each principal person involved.

3. Each operating member, as a condition of membership, will be required to sign the *CryptoSure Code of Conduct*; and to have membership re-validated, and the code re-signed, annually.

4. The *CryptoSure Foundation* will provide a *Dispute Arbitration Service*, capable of making binding judgments and awards, and e.g. rescinding membership if/where appropriate.

4.6   Conclusions: Blockchain *Versus* Trust—The Expert Issues in Disputes Over Crypto Assets

Given the unavailability (yet) of my proposed *CryptoSure* regime, the trust and reliability of 'blockchain only' systems and services as noted herein remain supposedly provided merely *technically*, by virtue of the 'distributed consensus' algorithm—that is, there is essentially and fundamentally no TTP involved or standing behind the creation and valuation of, and dealing and trading in, blockchain-held *Crypto Assets*.

The Internet is not a sue-able party. It has no intrinsic financial value, and 'belongs' to no-one. Since a Crypto Asset fundamentally consists of zeros and ones scratched on an Internet-accessed blockchain, changes stored and processed, written into, a distributed ledger, it may seem futile, perhaps legally meaningless, to ascribe a tangible value to a decentralised blockchain, without any substantive, sue-able TTP responsible for or standing behind its integrity and security.

However, when Crypto Assets become the subject of disputes—*Crypto Dragons*, as such disputes have been christened by me in recent articles published, for example, in *Solicitors Journal* and *The Expert Witness Journal*—the identification, location, and financial valuation of any Crypto Asset, access to it, holdings of it, and dealings and trading in it, will be critical.

And here's a key point: *Crypto Asset holdings and dealings are certainly not beyond legal protection or action, nor regulatory reach.* Although a Crypto Asset may essentially be 'decentralised digital vapour' a Court of Law can make a binding order to get forensic traction on it, because of the legally well-established *Obligation of Disclosure*. This obligation applies as much to a digital Crypto Asset as it does routinely to all other computer-held digital materials and 'documents', i.e. the Electronic Evidence relevant to any forensic investigation, whether for a Civil Dispute or for a Criminal Prosecution.

Thus, *Disclosure and Valuation of Digital Assets*, including Crypto Assets, is a significant issue arising in such financial and technology legal actions, civil or criminal. During years of expert witness work I have routinely assisted solicitors and Senior Counsel in framing appropriate technical *Requests for Disclosure*, and at request of attorneys I have drafted a *Checklist* giving practical, generally applicable wording for an effective *Digital Asset Disclosure* exercise. Details of my *Digital Asset Disclosure Wording Checklist* are summarised in my October 2019 article in *Solicitors Journal*.[10]

---

[10] 'Authored by AI—Here be crypto dragons: it's all about the evidence, proclaims the *Castell-GhostWriteBot*', Stephen Castell, *Solicitors Journal*, October 2019, pages 43–45.

https://www.solicitorsjournal.com/feature/201910/authored-ai.

'Can you tell if this has been authored by a robot? Would it matter, legally or otherwise, if you couldn't? Are you crypto-friendly, or if not, at least crypto-aware? …'.

And see:

'The future decisions of RoboJudge HHJ Arthur Ian Blockchain: Dread, delight or derision?', Castell, S. (2018), Computer Law & Security Review, Volume 34, Issue 4, August 2018, Pages 739–753, the Landmark 200th issue of CLSR under the Editorship of Emeritus Professor Steve Saxby.

The Checklist should assist litigation lawyers and ICT experts, in Financial Audit, Tax Assessment, Fraud and Theft Enquiry, Fintech Due Diligence, Investment Exchange Issues and Listings, M&A Projects, Corporate Risk Assessments, Divorce Proceedings, IP Conflicts and Smart Contract Audit forensic investigations.

More generally, some of the potential future issues that ICT systems professionals and experts may well be asked to investigate and upon which to provide analyses, conclusions and opinions, in regard to trust in, legal and technical reliability of, and associated disputes over, blockchain-based systems applications, are likely to include:

Cryptocurrency ICOs/ITOs

- Allegations of false or negligent representations in 'white papers', public issue documentation and presentations, websites.
- Failure to carry out due diligence as to project viability, systems and business integrity, quality standards, financial probity and implementation rigour.
- Consequential losses: investors losing money, businesses going bust and causality.

Blockchain

- Operational systems failures: the blockchain itself may be reasonably robust and reliable, but all interface/interconnect systems still need to be specified, designed, coded, constructed, tested and commissioned to acceptable ICT industry and professional quality assurance standards.
- Consequences: assessment of outages, denial, inaccuracy and unreliability of service, data transaction failures, errors or faults, data going missing, people losing money unable to conduct reliable business, smart contracts corrupted, distributed ledgers not capable of being trusted.
- Assessment and apportionment of causality, liability and responsibility for damages, losses and compensation.

Blockchain and GDPR Article 17

- In regard to requests 'to be forgotten' by data subjects, where their personally identifiable data are held on, or linked to, 'permanent and immutable' blockchain records: advice and management in regard to Court Orders granted for 'erasure'.
- Opinion as to efficacy of 'erasure' techniques, transactions, technologies, processes, proposed or implemented.
- Verification of the 'erasure' carried out: what constitutes sufficient evidence and proof of accuracy, correctness, completeness and persistence?

---

https://doi.org/10.1016/j.clsr.2018.05.011. While many are concerned about defining and developing AI Machine Ethics, *Castell's Second Dictum:* '*You cannot construct an algorithm that will reliably decide whether or not any algorithm is ethical*' (2017) reveals that this is a futile exercise. 'Talking about the ethics of machines might be like speaking of the happiness of water' (page 743).

'Revolution of securities law in the Internet Age: A review on equity crowd-funding', Tao Huang and Yuan Zhao, *Computer Law & Security Review*, 33, (2017) 802–810.

- Assistance with discussions with Information Commissioner's Office as to validity of requests 'to be forgotten', confirmation of the extent, reliability and security of 'erasure' (to be) carried out, and reasonableness of any possible/proposed fines or penalties to be imposed.

Ownership of IP

- Advice and guidance as to: whether relying on third party platforms, or developing its own blockchain software, any company seeking to build blockchain-based applications runs an IP infringement risk (there are no ISO standards, and more than 1000 blockchain patent applications filed with the US Patent Office).
- Assessment of impact, consequences, remediation: e.g. litigation over patents and software copyright.
- Expert investigation, search and advice as regards prior art, and/or lack of inventive step, for patent infringement actions and challenges to the original Grant of Patent.
- Advice and guidance in connection with negotiations with patent or copyright owners over use restrictions, licence fees and development capability.

This is of course in addition to the 'usual' relentless occurrence of disputes over computer systems failures generally. Failures of confidence and good faith (*Cambridge Analytica* private data misuse), of dependable cybersecurity (*Facebook* password hacking), of mission-critical financial systems (*TSB* online banking deficient systems upgrade), of product 'fitness for purpose' (*VW Dieselgate* emissions 'cheat' software), of clinical operational reliability (*NHS* faulty breast cancer-screening algorithm), and of aircraft flight systems reliability and integrity (Boeing 737 MAX crashes): these are just a few examples of a growing stream of ever-upscaling IT Disasters that have regularly emerged over the past thirty years.

I have been involved as expert witness in the largest and longest computer software and systems contractual disputes to date reaching the English High Court, and Sydney Supreme Court, with damages claimed in such actions in the hundreds of millions of pounds. Indeed, twenty years ago, in the USA *Foxmeyer* case, the failure of an entire substantial multi-billion corporation occurred and was directly due to the faulty implementation and management of a major company-wide computer systems upgrade project.[11]

---

[11] https://www.slideshare.net/shaunaksontakke/batch-25-it-erp.

'ERP Case Study—Failure case—FoxMeyer Case Shaunak Sontakke ... April 17, 2014 ... FoxMeyer was the fifth largest drug wholesaler in the United States (1995) with annual sales of about 5 billion US$ and daily shipments of over 500,000 items. ...FoxMeyer was driven to bankruptcy in 1996, and the trustee of FoxMeyer announced in 1998 that he is suing SAP, the ERP vendor, as well as Andersen Consulting, its SAP integrator, for $500 million each...'.

http://calleam.com/WTPF/?p=3508.

'FoxMeyer Drugs A $65 M investment in an Enterprise Resource Planning System (ERP) and new warehousing facilities results in the destruction of a $40B business. ... Delays in delivery and the failure to fully realise the business benefits results in the organisation being unable to profitably service contracts it had entered into. ... cash flow issues forced the company into Chapter 11 bankruptcy. The company that had been worth $40 B prior to the project was then sold off for just $80 M to rival McKesson Corp...'.

With blockchain-based distributed ledger, smart contract and cryptocurrency developments and systems becoming ever more established, Crypto Dragon disputes—whether civil or criminal (thefts, scams, frauds)—are certain to increase, and potentially cause increasingly widespread and relentlessly larger financial and other anxiety, consequences and damages.

When it is *your* Crypto Assets that are the ones under examination in pursuit of, or arising from, disputes, allegations, valuations, tax demands, thefts, systems failures, prosecutions or other forensic investigations you had better hope—absent my proposed new *CryptoSure* regime—that there is a TTP to be held responsible for disclosing the Electronic Evidence essential to your case, rather than rely on the 'trustless' digital cipher of the blockchain 'distributed consensus' mechanism itself to be of any practical or material human assistance.

Chapter 5
Decentralized Governance for Smart Contract Platform Enabling Mobile Lite Wallets Using a Proof-of-Stake Consensus Algorithm

Vipin Deval, Alex Norta, Patrick Dai, Neil Mahi, and Jordan Earls

Abstract Permissionless blockchain-enabled smart contracts that use the proof-of-stake (PoS) algorithm for validation of transactions have scalability and on-chain governance as a primary concern. We consider smart contracts that use lite mobile wallets with support of the unspent transaction output protocol (UTXO) and simple payment verification (SPV) to increase the number of transaction throughput. Furthermore, we investigate in this paper to which extent it is possible to change the parameters of the blockchain using smart contracts in hot parameter swaps at runtime for enabling continuous soft forks. The leading smart contract solution Ethereum does not have the functionality to change blockchain parameters and requires hard forks multiple times, and thus, Ethereum smart contracts are not scalable for large industrial applications. This whitepaper fills the gap by presenting the Qtum's decentralized-governance protocol (DGP) that aims for managing essential blockchain parameters using smart contracts for enabling frequent soft forks. Compared with the Ethereum alternative, we explore the Qtum utility advantages and current Qtum smart contract future development plans integrating dedicated x86 hardware development for industry case applications.

5.1 Introduction

Smart contracts are agreements in which contract participants communicate through negotiation and use consensus mechanisms to enforce the behavior of the parties in the blockchain system. Smart contracts are applicable in diverse domains such as,

V. Deval (✉) · A. Norta
Blockchain Technology Group, Department of Software Science, Tallinn University of Technology, Akadeemia tee 15A, 12816 Tallinn, Estonia
e-mail: vipin.deval@taltech.ee

A. Norta
e-mail: alexander.norta@taltech.ee

P. Dai · N. Mahi · J. Earls
Qtum Foundation, Singapore, Singapore
e-mail: foundation@qtum.org

e.g., financial technology [5], Internet of things (IoT) applications [16], and so on. A decentralized transaction validation, initially using so-called proof of work (PoW) [22] where computationally expensive cryptographic riddles must be solved, is an integral feature of smart contracts. A distributed public ledger called the blockchain, which tracks transaction events without having a trusted central authority, is the key technology that comprises smart contracts. With the launch of Bitcoin [15], a peer-to-peer (P2P) cryptocurrency and payment system that provide a restricted range of protocol layer operations, blockchain technology is spreading in popularity. Bitcoins PoW transaction validation is electricity-intensive and computationally costly [7].

Unlike bitcoins, the Turing-complete language solidity[1] is used in several smart contract systems that match JavaScript syntax and enactment goals, e.g., the Ethereum Virtual Machine [24]. The validation of transactions by PoW reduces scalability to the point that most industrial applications consider Ethereum to be impracticable. Scalability is the term that relates various quantitative metrics to each other in software engineering to express that even when the amount of users increases, the software still maintains a reasonable response time. Yet, in the case of blockchain, scalability refers to the number of transaction validations per second in a blockchain networks [9]. Qtum[2] is the first smart contract platform that uses so-called lite wallets to address the scalability issue as in the whitepaper version 1 [10]. The first unspent transaction outputs (UTXO) [8]-based smart contract system with a proof-of-stake (PoS) [18] consensus model where tokens are staked is one of Qtum's key objectives. The above implies that the next block's maker is chosen based on the cryptocurrency wealth owned. A simpler version would be instead of mining blocks; they are minted or forged where each new block grants an additional reward to the forger. In addition to transaction costs, the latter means the forger earns an interest in the total sum of funds at stake. Qtum is compatible with the ecosystems of bitcoin and Ethereum that aims to create a bitcoin variation with EVM compatibility. Qtum uses business use cases for a strategy containing mobile devices in search of a realistic design approach. The latter enables Qtum to encourage blockchain technology to a broad range of Internet users and thus, to decentralize the validation of PoS transactions. To further expand the smart contract platform capability of the blockchain, QTUM plans to implement the x86 high-performance virtual machine that allows for directly programming smart contracts with C, C++, rust, go, etc., and also supports other technologies that can be used in a x86 architecture. The x86 virtual machine solves problems in the EVM such as not supporting the standard library, generating too large bytecode, not supporting floating points, and difficulties with debugging the code.

Although smart contract systems such as Ethereum attract interest, widespread market adoption does not exist for the reasons discussed above. This paper tackles the gap by defining the Qtum framework for smart contract systems that answers the question of how to establish the decentralized governance frameworks on a smart contract platform to address crucial customer requirements for scalable smart contract systems. To establish a separation of concerns, we pose the following subquestions.

---

[1]http://solidity.readthedocs.io/en/develop/.

[2]https://qtum.org/en.

What are the basic requirement sets of Qtum blockchain for evolutionary governance mechanisms to enhance scalability? What is the dynamic structure of the evolutionary governance architecture that supports the improved scalability? What abstraction layer features process a translation between the EVM and specific bitcoin protocols to allow for abandoning the bottleneck Ethereum account model with lite wallets? With the latter, clients can connect to the full node without downloading the entire blockchain using simple payment verification (SPV) [10]. The decentralized governance enables soft forks that allow to roll over tokens and avoids hard forking, thus increasing the throughput of the number of transactions per second and reducing the network latency. This significant performance is achieved by scaling the Qtum blockchain with decentralized governance [9]. Note that this paper is based on an earlier version of the Qtum technical whitepaper [10] that comprises extended background information.

The remainder of this paper is structured as follows. First, Sect. 5.2 gives preliminaries about the decentralized governance protocols (DGP) in the Qtum blockchain. Section 5.3 focuses on concrete advantages of the Qtum framework for achieving distributed governance in comparison to related solutions. Section 5.4 discusses the extensions of the Ethereum Virtual Machine in Qtum for achieving scalability through governance. Section 5.5 presents the Qtum abstraction layer that allows for lite smart contract wallets on mobile devices. Section 5.6 discusses, on the one hand, the technical realization of the Qtum hard forking update and additionally describes the Qtum x86 virtual machine for decentralized governance. Finally, Sect. 5.7 concludes this whitepaper together with discussing limitations, open issues, and future development work.

5.2 Preliminaries for Qtum Decentralized Blockchain Governance

According to [14], EVM has shortcomings such as previous attacks on mismanaged exceptions and dependencies such as timestamps and transactions ordering. The account abstraction layer (AAL) is essential to Qtum's connection between the UTXO layer and smart contract layer with recent innovation in Qtum implementation. Through the AAL, the UTXO is converted into an account model for execution by different virtual machines, such as Ethereum's EVM or the x86 virtual machine. Furthermore, the virtual machine's account balance can be converted to UTXO via the AAL. This mechanism realizes the hierarchical design of the balance and smart contract platform, adopts UTXO of Bitcoin for liquidity, and natively supports the multisignature algorithm that is more secure as the balance for billing and transition.

For most cryptocurrencies, block producers receive block rewards and transaction fees. The main purposes of transaction costs are: (1) incentivize block producers to enter the transaction into the block and (2) block spam transactions on the blockchain network. The bitcoin transaction-fee design based on the PoW consensus algorithm is very effective as it requires payment of mining equipment and electricity costs to generate blocks [4, 23]. Still, with the PoS consensus algorithm, the cost of equipment and electricity is low [17]. If a mining participant has a large amount of coins, then that mining participant only needs to wait for a specific time to generate a block, although the probability of generating a block does follow. In the Turing-complete smart contracts environments, it is possible to use spam transactions at a relatively small cost to attack a PoS-based network. Therefore, Qtum modifies the reward mechanism of PoS 3.0 [10] to partially solve the useless stake problem caused by a spam attack. We term the new PoS mechanism as mutualized proof of stake (MPoS) discussed in Sect. 5.3.3.

The governance of blockchain is an essential issue in the sustainable development of a blockchain ecosystem with an important impact. Establishing a more effective blockchain network governance model is a challenge in the future development of the blockchain industry. For example, solving the problem of bitcoin network expansion more effectively, advantageous choices in different bitcoin expansion plans, and how to achieve this consensus, so as to avoid 1MB block, or 2MB block debates of bitcoin that take up to 2–3 years [12].[3] An open issue is also reaching a consensus to fork and reduce the impact of forking on the bitcoin ecosystem. To this end, Qtum designs the DGP, a new governance model for blockchain networks. The DGP is an on-chain governance protocol implemented on the Qtum blockchain that allows changing some consensus values through the means of on-chain proposals and voting system. After the approval of changes, the parameters are applied seamlessly without the need for a hard fork.

The essence of a smart contract is a piece of code. In theory, a Turing-complete smart contract can realize the protocol design of arbitrary complexity, even including the core protocol of the blockchain such as the consensus part, efficiency trade-off means efficiency versus energy consumption and costs, and security. For security reasons, Qtum's DGP protocol is only suitable for performing changes to specific parameters within the security scope, while taking a certain time limit for the effective validity of the parameters. Thus, managing the parameters of the blockchain network via smart contracts that are embedded in the creation zone and implementing a decentralized network autonomy mechanism yields an automatic upgrade and fast iteration of the blockchain network without worrying about soft fork and hard fork.

With the Qtum DGP protocol and mechanism, the blockchain network is adaptive to certain forks without the need to upgrade the core wallet, or upgrade process, while minimizing the impact of forks on the blockchain network, ecosystem, and users. An independent DGP smart contract controls each blockchain network parameter under management. DGP governance completes the decision to modify the blockchain network parameters by voting and also the voting participants; the voting decides the management and governance seats. Consequently, all nodes on the wallet and blockchain can periodically check the recursive length prefix (RLP) store to determine if a new blockchain network parameter needs to be changed. The RLP store is the encoded state data in the Merkle Patricia tree (MPT) at every Ethereum node.

---

[3]https://cointelegraph.com/explained/Bitcoin-block-size-explained.

After discussing the preliminary concepts about the on-chain governance, we next describe the basic components of the Qtum blockchain to achieve governance on the blockchain.

5.3 Qtum Basic Requirement Sets for Decentralized Governance

As Ethereum uses an account model for wallets, and it is necessary to download the entire blockchain, which does not allow for wallets to operate on mobile devices with limited storage capacity. Furthermore, PoW is a performance bottleneck for large-scale industrial applications of smart contracts on blockchains. Thus, we explain the Qtum approach for overcoming these shortcomings of Ethereum. The remainder is structured as follows. Section 5.3.1 compares the advantages of bitcoin UTXO versus the Ethereum account model. Next, Sect. 5.3.2 describes the consensus mechanisms for the Qtum blockchain and Sect. 5.3.3 presents Qtum's novel MPoS. Finally, Sect. 5.3.4 presents the details about the AAL of Qtum and EVM integration.

5.3.1 UTXO Versus Account Model

In the UTXO model, the unspent input bitcoins are used in the transactions that are consumed and new UTXOs are generated as transaction outputs after the execution of the transactions. The unspent outputs for transactions are generated and returned to the spender [1]. The above process transfers a certain amount of bitcoins between various private key owners and new UTXOs are spent and created in the transaction chain. The UTXO of a bitcoin transaction is unlocked by a private key to sign an updated transaction version. Miners generate bitcoins in the bitcoin network utilizing a mechanism called a coinbase transaction with no inputs. For transactions with a restricted set of operations,[4] bitcoin uses a scripting language. The scripting framework processed data in the bitcoin network using stacks (Main Stack and Alt Stack), an abstract data form based on the last in, first out (LIFO) concept.

In the bitcoin client, the developers use `isStandard()` function [1] to summarize the scripting types. Bitcoin clients support: Pay to Public Key Hash (P2PKH), Pay to Public Key (P2PK), MultiSignature (less than 15 private key signatures), Pay to Script Hash (P2SH), and `OP_RETURN` [25]. With these five standard scripting types, bitcoin clients can process complex payment logics. Besides that, a non-standard script can be created and executed if miners agree to encapsulate such a non-standard transaction.

---

[4]https://en.Bitcoin.it/wiki/script.

For example, using P2PKH for the process of script creation and execution, we assume paying 0.01BTC for bread in a bakery with the imaginary bitcoin address "Bread Address". The output of this transaction is:

```
OP_DUP OP_HASH160 <Bread Public Key Hash> OP_EQUAL OP_
CHECKSIG
```

The operation `OP_DUP` duplicates the top item in the stack. `OP_HASH160` returns a bitcoin address as top item. To establish ownership of a bitcoin, a bitcoin address is required in addition with a digital key and a digital signature. `OP_EQUAL` yields `TRUE` `(1)` if the top two items are exactly equal and otherwise `FALSE` `(0)`. Finally, `OP_CHECKSIG` produces a public key and signature together with a validation for the signature pertaining to hashed data of a transaction, returning `TRUE` if a match occurs.

The unlock script according to the lock script is:

```
<Bread Signature> <Bread Public Key>
```

The combined script with the above two:

```
<Bread Signature> <Bread Public Key> OP_DUP OP_HASH160
<Bread Public Key Hash> OP_EQUAL OP_CHECKSIG
```

The script combination's execution is valid only if the unlock script and the lock script have a predefined matching condition. The latter means that the Bread Signature must be signed with the matching private key of a legitimate Bread Address Signature, and the result is correct.

Unfortunately, the bitcoin scripting language is not a Turing-complete language, e.g., no loop feature. The latter is not a widely used language for smart contracts programming. Limitations eliminate safety risks by avoiding complex payment conditions such as creating endless loops, or other complicated logical loopholes.

In the UTXO model, the history of any transaction can be traced transparently via the public ledger. The UTXO model has the ability to initialize transactions between multiple addresses in parallel processing, suggesting extensibility. In addition, the UTXO model promotes privacy in that it is possible for users to use Change Address as a UTXO output. Qtum aims to incorporate smart contracts based on the groundbreaking UTXO concept.

Ethereum is an account-based system[5] as compared to the UTXO model. More specifically, every account encounters direct value and data transfers with state transitions. An Ethereum account address of 20 bytes contains a nonce as a counter to ensure that a transaction processed once, the balance of the main internal cryptofuel to pay transaction fees called Ether, optional contract code, and default-empty storage accounts. The two forms of Ether accounts are externally managed by a private key on the one side and contract code on the other. External accounts are used for the creation of information transmissions and transaction signatures. The contract account is used to create a contract or to send other information after receiving an internal storage read and write operations.

---

[5]https://github.com/ethereum/wiki/wiki/White-Paper.

In Ethereum, balance management is comparable to a real-world bank account. The global status of other accounts is theoretically affected by the newly created blocks. Each account has its own balance, storage, and code-space base to store related execution results for calling other accounts or addresses. Users perform P2P transactions via client remote procedure calls in the current Ethereum account scheme. While it is possible to send messages via smart contracts to more accounts, these internal transactions are only visible in the balance of each account, and it is a challenge to track them on the Ethereum public ledger.

Based on the above argument, we consider the Ethereum account model to be a bottleneck for scalability and see the bitcoin network UTXO model's substantial benefits. Since the latter strengthens the network effect we want to deliver, adopting the UTXO model is a critical design decision for the pending Qtum update.

5.3.2 Consensus Management

Consensus discussions are continuing for which platform meets the criteria of the respective projects. The most commonly discussed areas of the consensus are the following: POW [21], POS [2], Complex POS[6], and Byzantine Fault Tolerance [6], as HyperLedger discussed. The essence of the consensus is to achieve data consistency with the distributed algorithms. The available alternatives, for example, are the Fischer Lynch and Paterson theorem, which states, without the 100% agreement between nodes, the consensus cannot be achieved.

Miners participate in the bitcoin network in the hash collision verification process via PoW. If the miner is able to calculate the hash value and fulfills a specific condition, then the miner assert to the network that a new block is mined:

$$Hash(Header\,Blocks) \leq \frac{M}{D}$$

For the amount of miners M and the mining difficulty D, the Hash() represents the SHA256 power with value range [0, M], and D. The SHA256 algorithm used by bitcoin allows every node to quickly verify each block if the number of miners is high compared to the mining challenge.

The BlockHeader of 80 bytes differs with each Nonce, while the generic difficulty level changes dynamically with the blockchain network's total hash capacity. If two or more miners solve a block simultaneously, a small fork occurs in the network. This is the point where a decision is needed to accept or deny block by the blockchain. The chain that has the most proven work in the bitcoin network is legal.

Most PoS blockchains can source their heritage back to PeerCoin[7] that is based on an earlier version of Bitcoin Core. There are different PoW algorithms such as

---

[6]http://tinyurl.com/zxgayfr.

[7]https://peercoin.net/.

Scrypt,[8] X11,[9] Groestl,[10] Equihash [3], etc. Implementing a new algorithm aims to prevent the accumulation of computing power by one entity and prevents the introduction into the business of application-specific integrated circuits (ASICs). Qtum Core selects PoS for necessary consensus creation based on the latest bitcoin source code.

The generation of a new block in a typical PoS transaction must fulfill the following condition:

$$Proof\,Hash < coins \times age \times target$$

The stake modifier [20] computes together with the unspent outputs and the current time in `ProofHash`. This way, a malicious attacker can initiate a double-spending attack by collecting large quantities of the coin age. Another issue posed by coinage is that nodes are intermittently live after rewarding, rather than online continuously. In the enhanced version of the PoS agreement, the elimination of the coinage allows more nodes to be online simultaneously.

Due to the potential coinage attacks and other forms of attacks [13], the original implementation of POS suffers from many security problems. Qtum agrees with the Blackcoin team [20] security review and integrates PoS 3.0[11] into the most up-to-date Qtum core. Theoretically, PoS 3.0 rewards investors who *stake* their coins longer, but does not allow coin owners to leave their wallets offline.

5.3.3 Mutualized Proof-of-Stake

Qtum requires a more dynamic fee mechanism because of a focuses on smart contracts and blockchain calculations. Thus, transaction processing requires more time to interact and execute with smart contracts while there are several security risks in Qtum. For example, a danger is that an attacker executes a malicious program by paying an expensive fee. Since these costs are attributed to the block producer, the cost of the attacker will eventually become small. The cryptocurrency based on the existing PoS system is not affected by such attacks because of not supporting Turing's complete smart contract VM. Qtum is the first blockchain based on PoS and Ethereum's contract to implement the gas mechanism that regulates the computational uses and capabilities allowed on the blockchain.

Qtum further modifies PoSv3[12] to partially address the useless stake problem caused by spam attacks. Qtum uses PoSv3 and changes the payment system for block rewards and transaction fees. Compared to block producers who immediately

---

[8] https://litecoin.info/Scrypt.

[9] http://cryptorials.io/glossary/x11/.

[10] http://www.groestlcoin.org/about-groestlcoin/.

[11] http://blackcoin.co/.

[12] https://blog.qtum.org/qtums-proof-of-stake-consensus-at-a-high-level-ea53e051ac66.

receive block rewards and transaction fees, the new approach is to reward and trade costs shared among multiple miners in the network. Specifically, the block producer receives 1/10 of the total cost when generating the block and after receiving the predetermined block confirmation time, the block producer receives 1/10 of the cost of the other 9 blocks.

This new approach appears to be a minor change while it solves several security risks that affect Qtum. A considerable hidden danger is that an attacker may become a potential block producer by buying enough coins, rendering the network vulnerable to DoS attacks through certain EVM opcodes for a small price. Even a malicious transaction consumes transaction costs and gas and an attacker can execute a malicious program by paying an expensive fee. Since these costs are attributed to the block producer, the cost of the attacker eventually becomes very small. With MPoS, this is not possible since generating a block can only receive 1/10 gas, unless he can allocate additional 9 blocks. Otherwise the remaining 9/10 of the gas are lost to other miners on the network. As a result, this type of spam attack can become very expensive.

MPoS also has shortcomings that we can solve by combining several additional technologies. For example, a malicious block producer can choose to accept his own very low price tag transaction, thereby minimizing the currency spent on execution and attack. This can be avoided by dynamically adjusting the minimum gas price, which is achieved by the decentralized autonomous protocol (DGP) to control a variety of junk transactions and contracts while monitoring whether they are malicious.

5.3.4 AAL and EVM Integration

Qtum is designed based on the Bitcoin UTXO model while the Ethereum smart contract platform adopts the account model. The smart contract as an entity requires an identifier, which is the address of the contract for operating and managing smart contracts via this address. The AAL is added to the Qtum model to transform the UTXO model into an account model that can be executed by smart contracts as Fig. 5.1 shows.

Qtum is based on bitcoin's blockchain architecture and uses the UTXO-based model. Therefore, Qtum has a layer of account abstraction that converts the UTXO-based model to an EVM account-based interface. Note that an abstraction layer in computing is necessary to mask specifics of the implementation of specific features to separate concerns about enabling interoperability and platform independence. For smart contract programmers, the EVM account model is easy to use as operations are available that check the balance of the current contract and other blockchain contracts and transfer ETH to other contracts as money (data-attached money). The AAL implementation of these operations is consequently more complex than predicted in Fig. 5.1. The operators used in the AAL are:

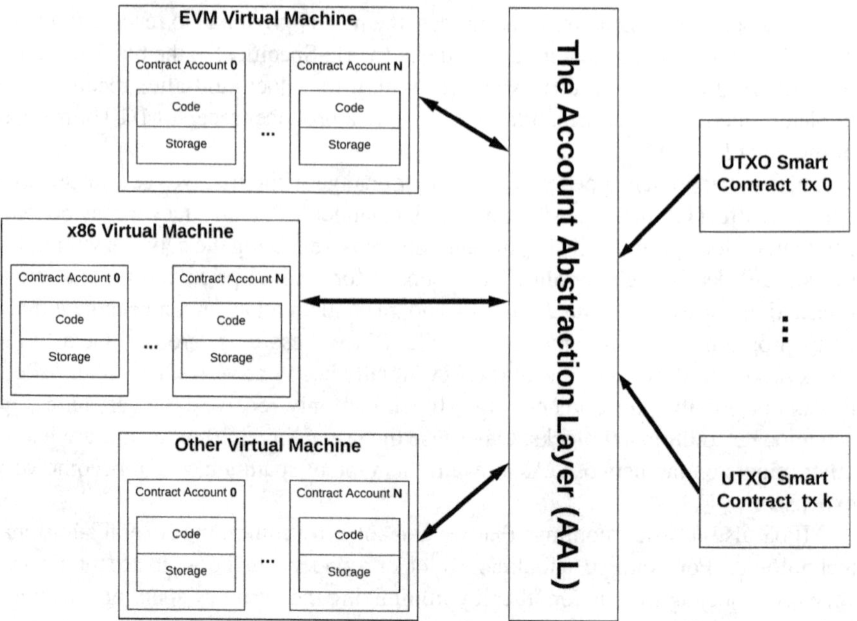

Fig. 5.1 The account abstraction layer

- **OP_CREATE**: Used to construct a smart contract, move the bytecode of the transaction to the contract-storage database of the virtual machine, and create a contract account.
- **OP_CALL**: Used to migrate relevant data and address information required for smart contract invoking and contract code execution. The operator can also send coins for smart contracts.
- **OP_SPEND**: The current contract hash ID is used for HASH input, or a HASH input transaction sent by UTXO to the contract, then the OP_SPEND is used for setting up a transaction script as an expense instruction.

In the bitcoin system, the corresponding transaction output can only be spent when ScriptSig and ScriptPubKey are verified. For example, ScriptPubKey usually locks a transaction output to a Bitcoin address (a hash of the public key). The execution script returns true only when ScriptSig matches ScriptPubKey (the system returns a value of 1) so that the corresponding transaction output is spent.

In Qtum, we emphasize the timeliness of smart contract execution. Thus, we add the OP_CREATE and OP_CALL operators to ScriptPubKey. When one of these opcodes is detected in a script, all network nodes start executing this script after a transaction has been put in a block. In this mode, the bitcoin's actual script language is less used as a scripting language and carries data instead to the EVM. The latter updates states within its own state database, upon execution by either of the opcodes, similar to an Ethereum contract. The EVM state changes in its own state database compared to the Ethereum implementation of smart contracts, contracts triggered by OP_CREATE, and OP_CALL operators.

Considered to use Qtum smart contracts easily, it is necessary to verify the data that triggers a smart contract and the public key hash of the data source. In order to prevent the proportion of UTXO on Qtum from being too large, we also design the transaction output of OP_CREATE and OP_CALL to be spent. Thus, the output of OP_CALL sends coins to other contracts, or public key hash addresses.

First, for smart contracts created on Qtum, the system generates a transaction hash for the contract call. The initial balance of the new contract is 0 (currently nonzero initial balance contracts are not supported). In order to meet the needs of the contract to send coins, Qtum uses the OP_CALL operator to create transaction output. Qtum defines the specific cost of a transaction (regardless of the out-of-gas case) as OutputValue, which is GasLimit. For the details of the specific gas mechanism, we refer to the reader the Qtum whitepaper 1.0 [10]. When the above script is added to the blockchain, the output is associated with the contract's account and is reflected in the balance of the contract, being the sum of the available contract costs.

After giving blockchain-oriented Qtum technicalities, we next describe the Qtum transaction processing and the architecture of the governance protocol.

5.4 Scalability Extension of the Virtual Machine Through Decentralized Governance

The EVM has a 256-bit machine word functioning using a stack and smart contracts that run on Ethereum use this virtual machine for execution. The EVM is designed for the Ethereum blockchain and assumes that an account-based approach is used for all value transfer. Thus, this section is structured as follows. Section 5.4.1 presents the Qtum transaction processing system and Sect. 5.4.2 details the Qtum DGP.

5.4.1 Qtum Transaction Processing Using Virtual Machine

Qtum is based on bitcoin's blockchain design and utilizes the UTXO-based model. Therefore, Qtum has an account abstraction layer that converts the model based on UTXO into an EVM account-based interface. Note that a program abstraction layer is instrumental in hiding the implementation details of specific features to create a separation of interoperability facilitation and application independence concerns. All transactions in Qtum, as in bitcoin, use the bitcoin scripting language. In Qtum however, there exist three new opcodes as described in Sect. 5.3.4.

Traditionally, scripts are executed only when an output is attempted to spend. For instance, while the script is on the blockchain with a regular public key hash transaction, execution and validation do not occur until a transaction input references the output. The transaction is valid at this point only if the input script (`ScriptSig`)

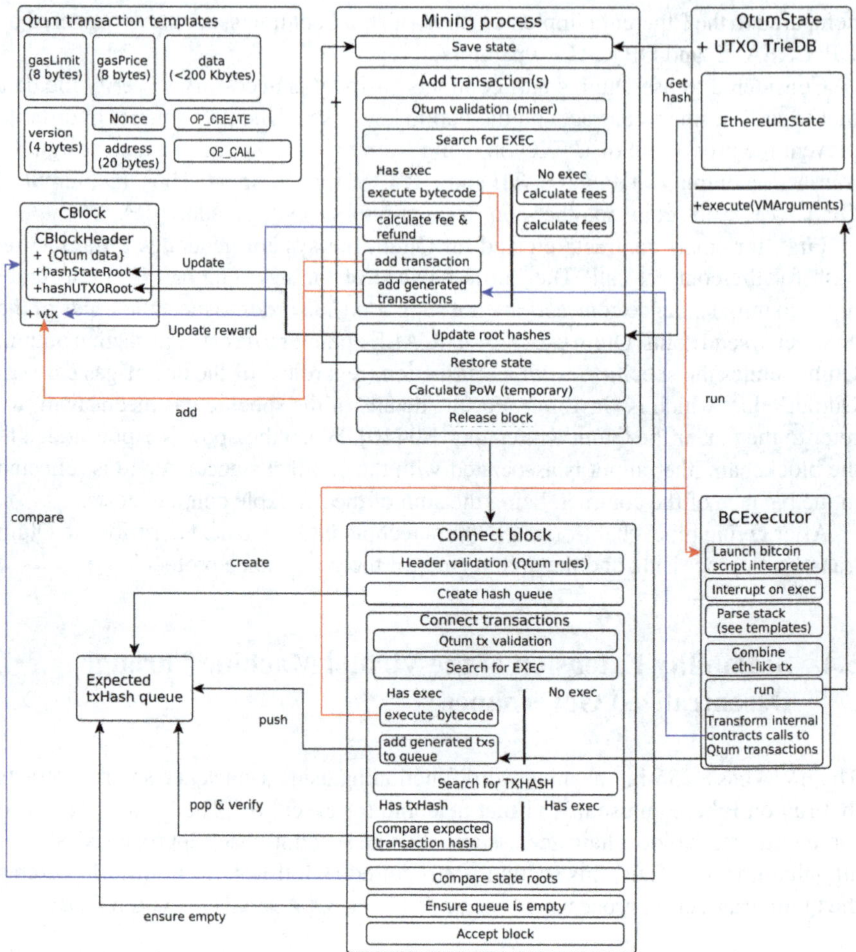

Fig. 5.2 Qtum transaction processing

provides the output script with accurate data that causes the output script to return nonzero.

Special processing of transaction output scripts (`ScriptPubKey`) containing either `OP_CREATE` or `OP_CALL` is done by Qtum, as shown in Fig. 5.2. When one of these opcodes is detected in a document, all network nodes execute it after the transaction is put in a block. The actual bitcoin script language in this mode acts less as a scripting language and carries data to the EVM instead. Upon execution by any of the opcodes, similar to an Ethereum contract, the latter changes state within its own state database.

We have to authenticate the data sent to a smart contract as well as its creator from a specific `pubkeyhash` address for easy use of Qtum smart contracts. The transaction outputs `OP_CREATE` and `OP_CALL` are also expendable to prevent the UTXO collection of the Qtum blockchain from being too large. The outputs of `OP_CALL` are spent on contracts when their code sends money to another contract or an address of `pubkeyhash`. Outputs from `OP_CREATE` are expended if the contract uses the operation `suicide` to delete itself from the blockchain.

5.4.2 Decentralised Governance Protocol

According to the recent Bitcoin Cash (BCH) fork process, the management of PoW chains such as bitcoin is oligopolized by large mine owners, affecting decentralization in the blockchain network [26]. Although Ethereum has community-based governance, it appears to be inefficient when reacting to sudden or urgent changes. At the same time, there is also the risk of "shadow government" [19]. EOS[13] is a smart contract platform designed to support large-scale applications and has improved the governance via centralizing the architecture and community constitution in which stakeholders vote for a delegate that governs the blockchain. Still, EOS fails to meet the governance needs of blockchain communities [11].

Therefore, a decentralized on-chain governance scheme is necessary as a structure for controlling and implementing changes to blockchains. Qtum DGP is an implemented on-chain governance protocol that allows changing specific consensus values through the means of an on-chain proposals and voting systems. Once the changes are approved and applied seamlessly, there is no need of a hard fork.

There are currently four DGP contracts deployed on the Qtum blockchain to control four values:

1. **Block size**: The maximum size of blocks (default is 2 Mb, DGP contract ID: 0000000000000000000000000000000000000081)
2. **Minimum Gas Price**: The minimum gas price accepted by consensus rules (default is 40 satoshis, DGP contract ID: 0000000000000000000000000000000000000082)
3. **Block gas limit**: The maximum gas that can be consumed in a block (default is 40 millions, DGP contract ID: 0000000000000000000000000000000000000084)
4. **EVM gas price schedule**: A schedule with each EVM opcode cost in gas (DGP contract ID: 0000000000000000000000000000000000000080).

The above four initial DGP controllable parameters are the most critical on the Qtum blockchain that may change. It is possible to add more parameters to be governed by the DGP with a hard fork of Qtum. That way, the changes in the consensus code can manage the new parameters being controlled by on-chain stored values.

DGP Features The key is to use the characteristics of the smart contract while writing the governance frameworks and rules into smart contracts that perform decisions

---

[13]https://eos.io/.

democratically. The DGP framework includes community processes, smart contracts (framework contracts, feature contracts), contract validation processes, management teams, and administrator. The framework contracts are responsible for electing the adminkeys and govkeys while the feature contracts are responsible for changing the parameters for the consensus value.

The management team refers to who can participate in management, role division, democracy, and concentration mechanism. The management team is called the management committee because all members (except the initial administrative staff) are generated by voting, including administrative staff and the governor. The administrative staff is responsible for the maintenance of the operational mechanism of the committee and also participates in the exercise of rights, that is, the voting of various proposals (including characteristic contracts). The operational mechanism of the administrator includes adding more administrators and governors as adminkeys and govkeys. The manager is responsible for the exercise of the committee's external rights, that is, the vote for the characteristic contract proposal.

The ideal DGP process has the following steps:

1. Proposal after the community discussion (characteristic smart contract)
2. The administrative committee of the management committee submits the proposal to the DGP framework contract and initiates the voting.
3. All members of the management committee vote on the proposal
4. If the proposal is passed, then the DGP framework contract is closed, and the feature contract comes into effect. After the specified number of blocks, new features of the feature contract are applied to the new blockchain.
5. If the vote does not pass within the predetermined time, the process ends.

Features of Qtum DGP:

1. **Smart Contracts**: DGP uses smart contracts to implement the ability to integrate smart contracts so that they can be exposed to rules, deployed in advance, and executed reliably.
2. **Multisignature DGP**: The management committee usually comprises multiple people. The current setting may have up to 30 members. Thus, the voting process that takes effect on the feature contract is the implementation of multisignature.
3. **Hot update, soft fork**: At present, Qtum's DGP is only a dynamic modification of the basic attributes of the blockchain. The feature contract only provides new online deployed attribute data and the DGP's effective process does not lead to a hard fork.

DGP Contract Function There are two types of DGP smart contracts. One is a framework contract that is responsible for completing the work of the management committee, authority management, and voting. The other one is the characteristic contract that is responsible for providing characteristic data.

The combination of the two types of contracts complete Qtum's chain management functions. We briefly introduce the functions of the two types of contracts: The framework contract completes the management and voting functions of the management committee. The personnel management includes the addition and deletion of

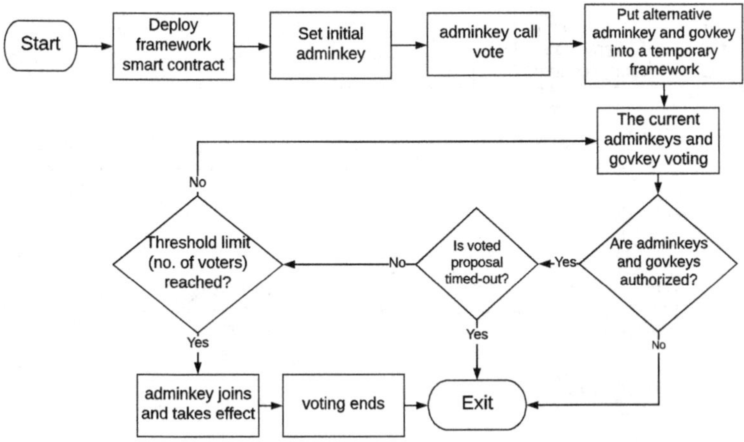

Fig. 5.3 Voting process to add the adminkeys and govkeys

two roles (admin and governor). The voting function includes the voting process and voting result management.

The functional definition of the framework contract comprises:

1. Proposal and voting, counting votes according to msg.sender.
2. Key Management (management of administrative and management personnel).
3. Management characteristic data (characteristic contract address).
4. Support for the modified rollback (not directly implemented, but can be achieved by revoting the feature contract before deployment).

DGP Implementation The functional implementation of DGP relies heavily on framework contracts. The implementation of the personnel-management function in the framework contract is exemplified by the process of adding the administrator as adminkey. As shown in Fig. 5.3, there exists a voting process to elect adminkeys and govkeys. The method of adding an executive as an adminkey is the same as adding a manager and the same holds for the process of deleting a person. The process in Fig. 5.3 is derived from the publicly available smart contract code for Qtum on GitHub.[14]

In addition to personnel management, the framework contract has a voting function. Before starting the voting, the feature contract must be deployed on the blockchain and then add the address of the feature contract to the framework contract that starts the voting. The corresponding flow is depicted in Fig. 5.4. The implementation of the DGP is concentrated on three functions: addAddressProposal, removeAddressProposal, and changeValueProposal. The specific implementation of the code is not described in detail in this paper and we refer interested readers to study the addAddressProposal function[15] carefully.

---

[14]https://github.com/qtumproject/.

[15]https://github.com/qtumproject/qtum-dgp/blob/master/dgp-template.sol.js.

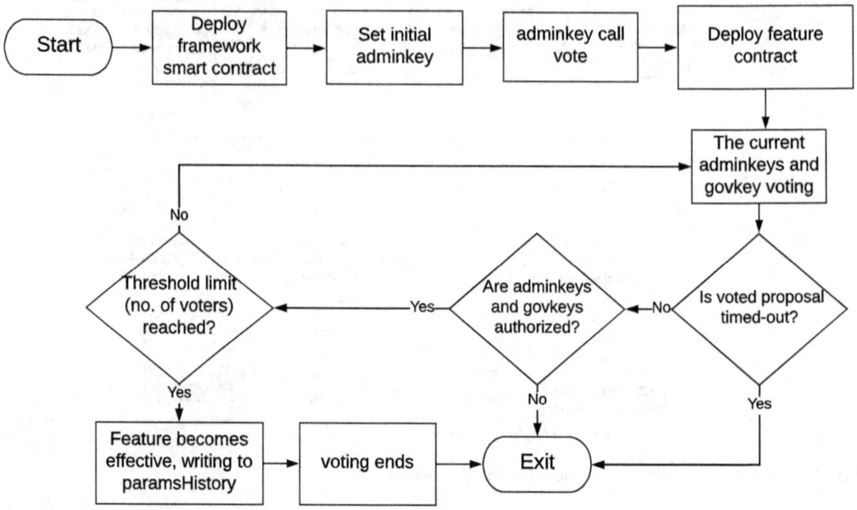

Fig. 5.4 DGP voting process to change the consensus parameters

Defects and Deficiencies Although Qtum's DGP implementation is simple enough with framework contracts and feature contracts, there are still flaws and shortcomings existant from the above functional implementation analysis. The DGP functionality relies on the default framework contract entirely and each framework contract only manages a single property. The current DGP framework has limited flexibility. Therefore, the current proposal for a framework contract is voted on.

Secondly, the DGP framework contract has no constructor as management parameters are not initialized and also not convenient to use. At the beginning of the deployment, the management committee only has one adminKey. Although the management parameters are set by directly calling the changeValueProposal interface, this is a voting process and it is not the right moment to initialize the voting process.

Third, the boundaries of role division are not clear. Although adminKey and govKey have a role distinction, adminKey can participate in all behaviors, including voting on feature contracts so that the govKey role is not highlighted. The adminKey is responsible only for the administrative affairs of the management committee, such as initiating voting and not participating in decision making. The govKey is responsible for all legislative processes, such as voting and thus, reflecting the result of the decision through public power.

Fourth, the framework contracts are used indefinitely. The paramHistory utilized to hold the historical value, is an array and the state of the contract is stored on the blockchain, used only to add elements without reducing. Still, this results in a higher cost of use.

Fifth, the end condition of the voting status is not reliable. If a vote is not within the specified time (number of blocks), then the minimum number of votes are not reached for that proposal. If no one continues to vote, the vote remains in the voting state and does not end automatically. A vote for a new feature contract cannot be initiated on the framework contract and it is required to use the original contract address and type to call the feature contract again. Therefore, timing functions are not implemented within smart contracts and external timing is required to ensure timing reliability.

There currently exists no systematic security guarantee and although some security designs are mentioned, such as delays in effect, DGP contracts are not allowed for certain addresses. Still, the mentioned security designs are not systematic and security assurance requires further investigations. For example, there is no security review mechanism for feature contracts to assure complex feature contracts guarantee safety, and so on.

After the architecture of the governance protocol in the Qtum blockchain, we next describe the Qtum abstraction layer to allow the EVM the execution of transactions on Qtum blockchain.

5.5　Qtum Account Abstraction Layer

The EVM is designed to run on a blockchain that has an account model. Qtum, however, is built on Bitcoin, uses a UTXO-based blockchain and includes an AAL enabling the EVM to run on the Qtum blockchain without any significant changes to the virtual machine and existing contracts with Ethereum. The EVM account model is easy for smart contract programmers to use. There are operations that verify the current contract balance and other blockchain contracts with transactions to transfer money (data attached) to other contracts. While these steps tend to be very simple and minimal, they are not easy to enforce in the Qtum blockchain based on UTXO. The AAL performance of these activities is, therefore, more complex than planned.

A smart contract deployed by the Qtum-blockchain is accepted and rejected by its address and includes a newly deployed contract balance set to zero. No protocol in Qtum enables a nonzero balance contract to be deployed. Instead, a transaction uses a OP_CALL opcode to transfer money to a contract.

The output script in the below example sends money to a contract:

```
1; the version of the VM
10000; gas limit for the transaction
100; gas price in Qtum satoshis
0xF012; data to send the contract
(usually using the Solidity ABI)
0x1452b22265803b201ac1f8bb25840cb70afe3303;
ripemd-160 hash of contract txid
OP_EXEC_ASSIGN
```

The above simple script gives the OP_CALL opcode transaction processing. The value given to the contract is OutputValue if there is no out-of-gas, or other exceptions. We address the detailed description of the process gas below. By adding

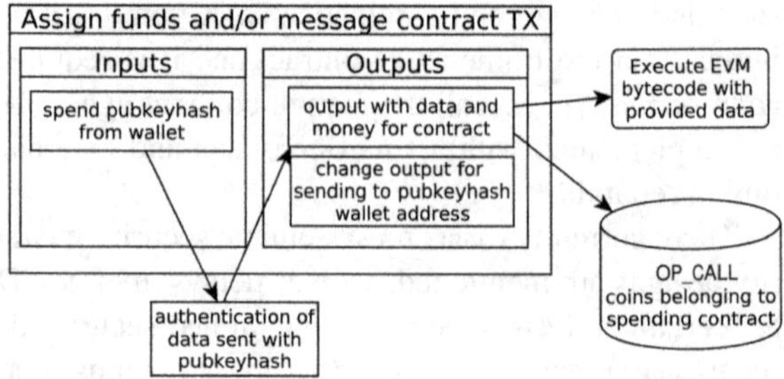

Fig. 5.5 Assign funds and/or a message contract TX

this output to the blockchain, the output is entered into the contract's UTXO domain. This production value represents the amount of expendable outputs in the balance of the contract.

Although Fig. 5.5 shows that money is sent from a standard public key hash output to a contract, the method of sending money from one contract to another is almost identical. The former spend one of its outputs when the contract transfers money to another contract, or public-key hash address. The sending contract requires the sending of the fund with planned contract transactions. The latter are unique since they must be in a block for the Qtum network to be legitimate. Expected contract transactions are created not by consumers but by miners during verification and execution of transactions. They are also not broadcast on the P2P network.

Planned contract transactions are primarily performed by a new opcode, OP_SPEND, in Fig. 5.6. The OP_CREATE and the OP_CALL internally have two distinct modes. The EVM is executed after the opcodes are executed as part of the processing of the output script. Still, when opcodes are executed in processing input files and EVM is not executed to prevent duplication. Instead, the opcodes OP_CREATE and OP_CALL behave comparable to no-ops and return either 1 or 0, that is, either spendable, or not spendable, depending on a certain hash transaction. Therefore, OP_SPEND is so important for this definition to work. Briefly, OP_SPEND is the latest opcode added that places SHA256 on the bitcoin script stack of the current spending transaction. The opcodes OP_CREATE and OP_CALL search for an expenditure test in the planned contract transaction list.

After the transaction passes to opcodes that appear in the Expected Contract Transaction List (usually from OP_SPEND), the result becomes 1, or can be spent. If not, the return is 0, or is not spendable. OP_CREATE and OP_CALL with vouts are therefore spent only when a contract is entered and therefore, the account abstraction layer requires that vout is spent, i.e., when the contract tries to send money. These results are safe and sound for contract funds to be expended only in compliance with a standard UTXO transaction under a contract.

For further details about the Qtum abstraction layer, we refer the reader to the Qtum whitepaper 1.0 [10].

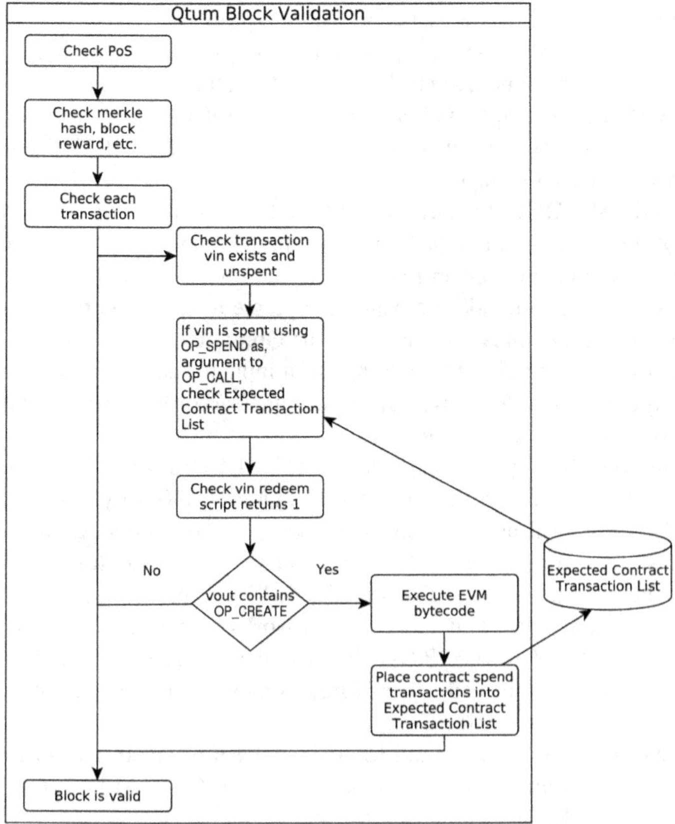

Fig. 5.6 Qtum block validation showing the expected contract transaction list

5.6 Discussion of the Technical Governance Realization

We present first in Sect. 5.6.1 the details about the Qtum hard forking process. Next, Sect. 5.6.2 shows the specific x86 virtual machine details that enable the execution of DGP smart contracts.

5.6.1 Hard fork Update

Every computer software needs to be updated regularly and in the blockchain domain, software updating is called hard fork after which, a blockchain performs according to a new set of rules. Therefore, Qtum core updates periodically by Qtum Improvement Proposals (QIP) according to proposals generated by the developers, the community users and the enterprise users. Qtum first releases the proposal to update the Qtum blockchain and if most of the users agree on the proposal, then a hard fork update occurs in the Qtum core. Generally, Qtum has many QIPs such as Qtum OP_SENDER opcode, difficulty adjustment algorithms, virtual machine upgrades and many more. For a more detailed description of various QIPs, we refer interested readers to Qtum GitHub.[16] From the available QIPs, below we discuss the Qtum OP_SENDER opcode to detect who sends a message.

Qtum OP_SENDER opcode: The blockchain platform for smart contracts, whether Qtum, or Ethereum, maintains a gas model to avoid a system crash caused by the exhaustion of blockchain resources because of smart contract use. Yet, the problem occurs that if an address wants to operate a smart contract, gas payment and address identification is necessary. For the Qtum blockchain, contract operators are judged by checking UTXOs as transaction inputs. Thus, an address must have Qtum as gas to make contract invocations, which poses difficulties to ordinary smart contract users and exchange institutions.

As a solution, QIP-5 proposes to add the OP_SENDER opcode to Qtum so that an address can invoke a smart contract without Qtum UTXOs by other users paying for Gas. This proposal adds the signature proof to the output script for contract transactions. The proposal also allows the user to sign the contract and prove the identity to the contract with the absence of QTUM, i.e., invoking the contract as the sender without paying gas. This proposal only works for contract calls, which allows an address without QTUM (UTXOs) to be set as the msg.sender. Still, the contract is not invoked for free and instead, other addresses must pay the corresponding QTUM consumption for the transaction.

Specification: A new opcode named OP_SENDER is added. This opcode is only valid when used within vouts that contain either an OP_CALL or OP_CREATE opcode. OP_SENDER takes three arguments:

- UniversalAddress type—the type of sender
- UniversalAddress data—the address data for the sender (for a dynamic-length account)
- scriptSig—the serialized scriptSig necessary to complete the signature of the address

Example vout

```
1 // pubkeyhash address type
address // pubkeyhash address
{signature, pubkey} // serialized scriptSig pushed
OP_SENDER
1 //EVM version
10 // gas price
100000 // gas limit
1234 // contract data to send
OP_CALL
```

---

[16]https://github.com/qtumproject.

OP_SENDER causes an internal script execution and fails if the result is not 1. For this example, the result is as follows: Example vout

```
signature
pubkey
OP_DUP
OP_HASH160
address
OP_EQUALVERIFY
OP_CHECKSIG
```

The serialized scriptSig is treated as if it were a normal scriptSig for validation purposes. If the result evaluates to 0, the transaction is invalid and cannot be included in a block. For pay-to-scripthash senders, the scriptSig is treated the same as normally, with the script being executed twice, first to check the hash is the same as the scripthash and then again to execute the redeemScript for spending P2SH outputs.

OP_SENDER Signatures: Normally in Bitcoin and Qtum, only scriptSig scripts within the vin contain signatures and thus, a signing scheme is available to avoid the signing signed-data problem. Still, this scheme breaks that assumption by having signatures inside of vout scripts and the signing split into two separate steps avoids bitcoin-based signature paradigms:

- Sign vouts containing OP_SENDER
- Sign vins

The signature process of vins remains the same and the vouts signature also is implemented by a signature method similar to Segwit.[17] This design may not be perfect, while for existing Bitcoin and Qtum blockchain networks, this design is more secure and forward compatible. In the scheme proposed by QIP-5, the signature of vin is not affected, vout is signed in Segwit mode and the OP_SENDER opcode is introduced into the transaction. By constructing the transaction with OP_SENDER, the address without QTUM UTXOs can invoke the smart contracts in a way where others perform gas payments.

5.6.2 Qtum X86 Virtual Machine

Originally, Qtum intents to support a variety of virtual machines and Qtum-x86 is the latest prototype that introduces the x86 virtual machine. Currently, it is possible to write smart contracts in C, while more languages will be supported in the future. This prototype is strictly a preview, and the contract interface is subject to changes before the final release. We design a new x86 virtual machine for which the Ethereum Virtual Machine is the first version. The x86 virtual machine uses the von Neumann computer architecture, i.e., the code is data in line with contemporary

---

[17]https://www.coindesk.com/information/what-is-segwit.

mainstream programming models. The basics of the x86 virtual machine ensure that smart contracts running on Qtum can be written with a few simple modifications and many existing compilers, or programming languages. Currently, almost all compilers support the x86 architecture instruction set, rendering the actual bytecode and architectural support as quite complete where the x86 virtual machine environment development is similar to the embedded development. In the development process, we need to pay attention to safety and try to remove safety hazards.

This version of Qtum comprises the following restrictions:

1. Must be used from Docker, or compiled from source.
2. EVM contract usage is not supported.
3. The testnet and mainnet networks cannot be used and will not sync as only regtest is supported.
4. Orphaning blocks and reorganizing the blockchain does not work properly.
5. Only the command line RPC interface is supported for x86 contracts, though the GUI still function correctly.

x86 Instruction Set: The virtual machine implements the i386 instruction set and subsequently expands to support the i686 instruction set. Since a smart contract is mainly an application layer program and for security reasons, only the ring 3 level and some necessary ring 0 instruction sets are supported. For the support of multiple programming languages, the significant languages for x86 virtual machines include C, C++, and Rust. We choose C and C++ because they are generally relatively simple, and we select Rust because it is relatively lightweight. The design concept of Rust pays special attention to security and prevents bugs. We first implement a GCC toolchain prototype that supports C, C++, and the very basic libC library. In the future, Qtum will develop an x86 virtual machine for Go and Python.

x86 Memory Model: x86 uses linear memory space and supports up to 4GB memory size. The ample memory space and its efficient operation code set enable complete blockchain data for smart contract analysis that is impossible on Ethereum Virtual Machines. In the future, we support AI-based smart contracts to monitor the blockchain automatically and thus, becoming a potential oracle, allowing smart contracts to dynamically adjust themselves to run as efficiently as possible under current network conditions. These blockchain data can include complete transaction data as well as node statistics (consensus correlation) that may be publicly available since these data are constants and require very little memory space.

Simplified State Storage and Consistency-Check Methods: The x86 virtual machine implements a new contract state storage method. For EVM, storage is a challenge, e.g., if multiple flags exist, alternatives must be found to package, or use more complicated operations because of 256-bit or 32-byte limits. x86 uses a new database to store x86 smart contract state data with large storage capacity. The key or value has no data size limit, which means that it is possible to use the key of any byte and point it to the variable value of the same length. This key saves the storage of state trees compared to Ethereum as the MPT of Ethereum stores the state index and its content information at all contract addresses of the current block. The index needs to be serialized to disk and the total amount of this data is significant. As the number of blocks increases, the entire node of Ethereum requires vast storage space to store state index information for all blocks. Moreover, Ethereum has to use the Bloom filter to acquire the state information from a very large database. The Bloom filter itself requires additional maintenance to work correctly.

x86 Standard Library: The EVM lacks a standard library and therefore, creates a challenge for the developers. The standard library renders the Qtum blockchain more efficient and also provides a unique internal code for these standard library functions, similar to Ethereum's precompiled contracts. This feature does not require special support for new precompiled contracts. Conversely, contracts may use the same unoptimized code that can be optimized as needed without any impact on the consensus. QtumOS is the contract execution initialized code generated when the contract is compiled. The user's smart contract establishes the execution environment through QtumOS. The standard library is the compiler's built-in support library that is similar to the one of the operating system. Therefore, Qtum needs a management mechanism to render the ecosystem more efficient and the Gas model for these library functions may need to be adjusted to reflect its true resource costs. As future work, Qtum will also introduce the concept of DGP to manage the standard library.

Trusted Library: A trusted library comprises trusted programs implemented by contract calls. Trust means, after a public test to ensure reliability, a rapid defect-remediation mechanism is designed through the DGP to minimize losses. This mechanism tackles problems such as the loss of funds caused by the call library problem exposed by Ethereum and avoids unnecessary hard forks. Additionally, the trusted library accelerates certain functions with minimum gas set at a fixed value. That way, smart contract developers can browse a list of trusted libraries that are precertified and optimized instead of implementing, or deploying additional code themselves, or paying for extra code. The trusted libraries encourage smart contract developers to select a trusted library for developing their programs and achieve better security.

Toolchain Setup Currently, there are no compiled binaries provided for Qtum-x86 since code changes happen too often to gain any value from setting up a binary compilation and distribution process. Thus, currently, the toolchain and Qtum-x86 itself must be compiled from source. A dockerfile is provided to simplify the build process that is otherwise quite complex with several dependencies and OS-specific steps. The Docker file and some utilities can be downloaded from the link.[18]

New Features in Qtum-X86 Qtum-x86 implements many features not supported for neither Ethereum's EVM nor Qtum's version of EVM. Many include features that expose the full power of the Qtum AAL for a more natural interaction of contracts with the UTXO model of Qtum. Other features are those more natural in x86, or that are very difficult to implement with the EVM's design.

- **UTXO Interaction**: Qtum-x86 contracts can directly interact with the various pieces of a transaction. The contract, in this case, reads the address (or addresses) that spent coins for them to be sent to another address. The contract logs this action, or does other processing, such as automatically releasing XRC20 tokens to the sender. The above processing is possible in an EVM contract in a more

---

[18]https://github.com/qtumproject/qtum-docker/tree/master/proto-x86.

complicated way, while sending to different address types requires special care and otherwise abnormal transactions.

- **Contract Events**: One of the most significant iterative improvements over the EVM is the implementation of contract events in Qtum-x86. Events ("logs" in Ethereum) are bits of data that are not saved directly into the blockchain but can be proven to exist within it. In Ethereum, this data is in a strict format comprising up to four 256 bit "topics" and one 256 bit "data" field. Thus, 256-bit hashes are typically used for any data larger than 256 bits. In Qtum-x86, events are flexible, dynamic length key-value data with a helpful type ID for both fields. Thus, the contracts create data that can be easily read in its intended presentation.
- **Contract Upgrades**: Contract upgrades are trivial in Qtum-x86 compared to Solidity/EVM contracts and modifiable bytecode within the contract itself. Therefore, contract upgrades eliminate the need for the infamously tedious proxy contract pattern in Solidity. There is no need to split state, address, and bytecode into separate contracts. If the contract upgrade is minimal, it might be possible to patch only a single area of bytecode.

  The other contributing factor to easier contract upgrades is the explicit handling of states. In the EVM, both key and value data in the contract state are constricted to a 256-byte limit. It requires the Solidity compiler to automatically generate the actual key data used from a Solidity contract. In Qtum-x86, storage is managed explicitly by contracts without a variable abstraction. The dynamic length of key and value data allows for complex structures to be stored as either a key, or a value and thus, allowing easy direct usage of Qtum-x86 storage. It also allows for easier data migration into different formats.

  The key data restriction in the EVM also contributes to a security problem not present in Qtum-x86. In the EVM, if some storage-based array variable can be indexed to arbitrary values, it is possible to read or write to any variable in the 256 bit storage key space. This vulnerability does not exist in Qtum and thus, it is safe to allow for dynamic storage-array indexing.
- **Universal Address System**: In Qtum, there are many different types of addresses. Thus, Qtum is strictly different from Ethereum where only a single type of address exists that may be either a contract, or public key depending on the blockchain state.

  In Qtum, there are several types:

  - Pay to pubkey
  - Pay to pubkeyhash
  - Pay to scripthash
  - Pay to witness pubkeyhash
  - Pay to witness scripthash
  - EVM contract
  - x86 contract

In Qtum-x86, we introduce the concept of a UniversalAddress that encodes both type and data and is intended to be of the consistent format in mainnet, testnet, regtest and even forks of the Qtum blockchain. The UniversalAddress is strictly different from the Base58 encoding of addresses in Qtum where the type also doubles as an easy to read prefix for the user. Thus, each blockchain type uses different prefixes, while the UniversalAddress, on the other hand, is intended only for smart contract usage and should be mostly invisible to end-users. Currently, in Qtum-x86, the data portion of this address is a fixed size, while in the future, it will be of dynamic size.

- **External Contract State Access**: In Qtum-x86, it is possible to read directly from an external contract's state. There is no need for getter functions as in Solidity/EVM contracts. Instead, it is expected that each contract publishes a public state API consisting of the various keys guaranteed to stay in the same format in the contract, even persisting through contract bytecode upgrades.

5.7 Conclusions

This paper presents a decentralized governance strategy to achieve on-chain governance for scalable smart contract solutions. We describe explicitly how the Qtum DGP achieves the governance to manage the blockchain parameters. We also discussed the Qtum transaction-processing system that supports the DGP smart contract architecture to enhance the scalability of the Qtum blockchain. Furthermore, the Qtum account abstraction layer connects the UTXO layer and EVM smart contract layer. Via the AAL, the UTXO model overcomes limitations of the Ethereum account model that can be executed by different virtual machines, such as the Ethereum EVM, or x86 virtual machine. The first integration of EVM and Bitcoin UTXO protocol is achieved by the AAL. In the end, we provide the Qtum x86 virtual machine that allows to write smart contracts in C, Rust, go, or a C++ environment and run smart contracts on Qtum blockchain.

Qtum provides a new governance model for blockchain networks by designing a DGP. The latter manages the parameters of the blockchain network via smart contracts embedded in the creation zone and implements a decentralized network autonomy mechanism to achieve automatic upgrades and fast iterations of the blockchain network to avoid a detrimental impacts of soft- and hard forks on the blockchain network and user community.

To expand the smart contract platform capability, we have implemented an x86 high-performance virtual machine that allows to write smart contracts with multiple languages and be executed on the x86 architecture. The virtual machine supports the x86 instruction set, simplifies the state storage and consistency-check methods. The x86 standard library and a trusted library render the development and deployment of smart contracts more flexible and secure.

In summary, smart contracts are sociotechnical tools and thus, a wide range of commercial applications must be taken into account. Qtum's various real-life application cases are substantial evidence for the increases acceptance in industry. Additionally, Qtum's mobile-oriented development strategy is designed to support transaction processing with a highly distributed equity-consensus mechanism. At the same time, Qtum also considers the front-end user experience and incorporates application layer development into smart contract lifecycle management, which has otherwise not received sufficient attention in existing blockchain solutions yet.

References

1. Antonopoulos, A.M.: Mastering bitcoins (2014)
2. Bentov, I., Gabizon, A., Mizrahi, A.: Cryptocurrencies Without Proof of Work, pp. 142–157. Springer, Berlin, Heidelberg (2016)
3. Biryukov, A., Khovratovich, D., Equihash: Asymmetric proof-of-work based on the generalized birthday problem. In: Proceedings of NDSS'16, pp. 21–24: San Diego, CA, USA (February 2016). ISBN 1-891562-41-X (2016)
4. Borge, M., Kokoris-Kogias, E., Jovanovic, P., Gasser, L., Gailly, N., Ford, B.: Proof-of-personhood: Redemocratizing permissionless cryptocurrencies. In: 2017 IEEE European Symposium on Security and Privacy Workshops (EuroS&PW), pp. 23–26. IEEE, New York (2017)
5. Bussmann, O.: The Future of Finance: FinTech, Tech Disruption, and Orchestrating Innovation, pp. 473–486. Springer International Publishing, Cham (2017)
6. Cachin, C.: Architecture of the hyperledger blockchain fabric. In: Workshop on Distributed Cryptocurrencies and Consensus Ledgers (2016)
7. Chatterjee, K., Goharshady, A.K., Pourdamghani, A.: Hybrid mining: exploiting blockchain's computational power for distributed problem solving. In: Proceedings of the 34th ACM/SIGAPP Symposium on Applied Computing, pp. 374–381. ACM (2019)
8. Christidis, K., Devetsikiotis, M.: Blockchains and smart contracts for the internet of things. IEEE Access **4**, 2292–2303 (2016)
9. Croman, K., Decker, C., Eyal, I., Gencer, A.E., Juels, A., Kosba, A., Miller, A., Saxena, P., Shi, E., irer, E.G., et al.: On scaling decentralized blockchains. In: International Conference on Financial Cryptography and Data Security, pp. 106–125. Springer, Berlin (2016)
10. Dai, P., Mahi, N., Earls, J., Norta, A.: Smart-contract value-transfer protocols on a distributed mobile application platform. https://qtum.org/uploads/files/cf6d69348ca50dd985b60425ccf282f3.pdf (2017)
11. Ferreira, D., Li, J., Nikolowa, R.: Corporate capture of blockchain governance. Available at SSRN 3320437 (2019)
12. Göbel, J., Krzesinski, A.E.: Increased block size and bitcoin blockchain dynamics. In: 2017 27th International Telecommunication Networks and Applications Conference (ITNAC), pp. 1–6. IEEE, New York (2017)
13. Kiayias, A., Konstantinou, I., Russell, A., David, B., Oliynykov, R.: A provably secure proof-of-stake blockchain protocol (2016)
14. Luu, L., Chu, D.H., Olickel, H., Saxena, P., Hobor, A.: Making smart contracts smarter. In: Proceedings of the 2016 ACM SIGSAC Conference on Computer and Communications Security, pp. 254–269. ACM (2016)
15. Nakamoto, S.: Bitcoin: A peer-to-peer electronic cash system. Consulted **1**(2012), 28 (2008)
16. Ouaddah, A., Elkalam, A.A., Ouahman, A.A.: Towards a Novel Privacy-Preserving Access Control Model Based on Blockchain Technology in IoT, pp. 523–533. Springer International Publishing, Cham (2017)
17. Saleh, F.: Blockchain without waste: Proof-of-stake. Available at SSRN 3183935 (2019)
18. Serguei, P.: A probabilistic analysis of the NXT forging algorithm. Ledger **1**, 69–83 (2016)
19. Shermin, V.: Disrupting governance with blockchains and smart contracts. Strategic Change **26**(5), 499–509 (2017)
20. Vasin, P.: Blackcoin's proof-of-stake protocol v2 (2014)

21. Vukolić, M.: The quest for scalable blockchain fabric: Proof-of-work vs. BFT replication. In: International Workshop on Open Problems in Network Security, pp. 112–125. Springer, Berlin (2015)
22. Vukolić, M.: The Quest for Scalable Blockchain Fabric: Proof-of-Work vs. BFT Replication, pp. 112–125. Springer International Publishing, Cham (2016)
23. Wang, W., Hoang, D.T., Hu, P., Xiong, Z., Niyato, D., Wang, P., Wen, Y., Kim, D.I.: A survey on consensus mechanisms and mining strategy management in blockchain networks. IEEE Access **7**, 22328–22370 (2019)
24. Wood, G.: Ethereum: A secure decentralised generalised transaction ledger. Ethereum Project Yellow Paper (2014)
25. Xu, X., Weber, I., Staples, M.: Architecture for Blockchain Applications. Springer, Berlin (2019)
26. Zook, M.A., Blankenship, J.: New spaces of disruption? The failures of bitcoin and the rhetorical power of algorithmic governance. Geoforum **96**, 248–255 (2018)

Chapter 6
Blockchain Technology:
Security and Privacy Issues

Nathan Clark, Leandros Maglaras, Ioanna Kantzavelou, Nestoras Chouliaras, and Mohamed Amine Ferrag

Abstract Distributed Ledger and Blockchain technology decentralise the way we store data and manage information. Blockchain technology is the backbone of most cryptocurrencies, but also is being explored in other serious fields, such as new governance systems with more participatory decision-making, and decentralised autonomous organisations that can operate without human intervention. In all these cases, data immutability and security are highly valuable. Blockchain technology prevents data feed failure and corruption and creates mechanisms that grant protection to distributed systems.

6.1 Introduction

Centralised Ledgers are currently adopted by many companies, which involves placing databases at a fixed logical or geographical centres. These are often recognised as points of failure for many different reasons. Distributed Ledgers aim to remove

N. Clark · L. Maglaras (✉)
School of Computer Science and Informatics,
De Montfort University Leicester, Leicester LE1 9BH, UK
e-mail: leandros.maglaras@dmu.ac.uk

N. Clark
e-mail: p16036199@my365.dmu.ac.uk

I. Kantzavelou · N. Chouliaras
Department of Informatics and Computer Engineering, University of West Attica, Attica, Greece
e-mail: ikantz@uniwa.gr

N. Chouliaras
e-mail: nchouliaras@uniwa.gr

M. A. Ferrag
Department of Computer Science, Guelma University, Guelma 24000, Algeria
e-mail: ferrag.mohamedamine@univ-guelma.dz

Networks and Systems Laboratory, Badji Mokhtar University, Annaba 23000, Algeria centrality, and related issues, by ensuring that databases persist across entities and are regularly updated, for the purposes of verification and auditability [1].

Blockchain is a type of Distributed Ledger and is the technology designed to capture, verify, and store data in decentralised databases [1]. An instance of Blockchain can exist anywhere for any use case or application; however, the underlying principles remain the same; the contents of Blockchain databases can be made publicly accessible; however, the Blockchain only enables trusted associated nodes to verify and append data to the Blockchain [2].

In its basic structure, Blockchain is a chain of connected blocks, stored linearly and chronologically, with each block representing a record. These define the data and data participants but are also used to hold unique data signatures for the current and previous block [2]. Storing these cryptographic signatures means that changes to block information can be detected by newer blocks in the chain. By adopting this approach, Blockchain is able to deliver verified data with high integrity, meaning that record modification is challenging, and deletion is impossible, across its distributed structure [2]. This makes it suitable for an array of secure applications.

The purpose of this chapter is to investigate Blockchain technology, presenting the fundamentals and several associated applications. The technology will be investigated to determine whether the benefits overshadow the emerging limitations and risks, which is necessary to assess the future of Blockchain technology in its current form.

6.2 Distributed Ledger Technology

A ledger is described as a record of each financial transaction that takes place during the life of an operating company. Due to the nature of ledgers, most applications have traditionally followed a centralised approach with databases established in fixed positions. Whilst this methodology does provide companies with control over data and third-party involvement, it can be considered as a single point of failure with the potential to threaten the anonymity, privacy, and security of data. Distributed Ledgers were conceptualised to resolve such issues [3].

A Distributed Ledger does not have a central authority and instead adopts a database that spans across many entities or nodes Fig. 6.1; each is individually responsible for ensuring that their copy of the ledger remains updated when new data is committed to the Distributed Ledger [1]. In this way, if any node attempts to update the ledger, the other associated nodes will first attempt to verify the data using a consensus algorithm [4] this is the automatic process of nodes voting and agreeing on the new copy of the Distributed Ledger.

By applying a distributed approach, the transparency of data is significantly increased among nodes. This is because all data stored in the ledger have verifiable and auditable history, supported by the computation of cryptographic signatures using hashing algorithms. In this regard, SHA-256 is one type of hashing algorithm commonly utilised, much like Blockchain is one type of Distributed Ledger technology.

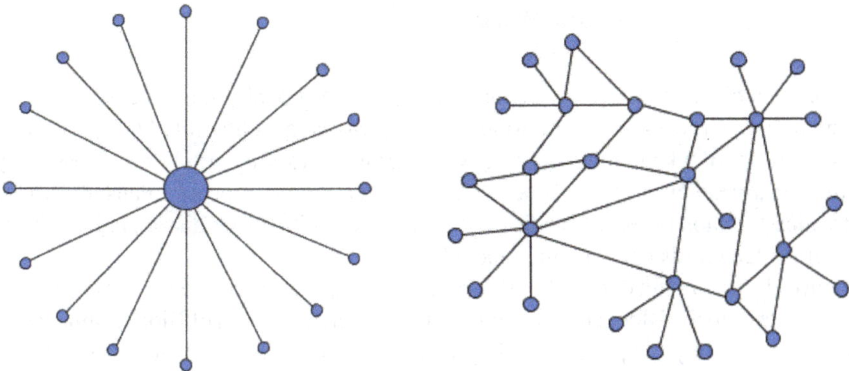

Fig. 6.1 Transition from a centralised to a distributed architecture

6.2.1 Blockchain Benefits

The idea behind Blockchain technology was first proposed in the early 1990s by two researchers, Stuart Haber and W. Scott Stornetta, as they sought to find a solution to digital file timestamp integrity. Their research lead to them developing a cryptographic chain of blocks which enabled document timestamp anti-tampering; this was later upgraded by Haber and Stornetta to improve efficiency. Despite their best efforts, this innovative technology was never adopted, leading to the patent lapsing in 2004. The technology was revolutionary for its time though not properly appreciated, possibly due to the other technological advancements being made, like the advent of the internet, the Linux operating system, among others.

Nevertheless, the theory of Blockchain was redefined in 2008 with a whitepaper titled "Bitcoin: A Peer-to-Peer Electronic Cash System", published by Satoshi Nakamoto (Nakamoto 2008). This paper discussed the issues present in commerce, like the need for trusted third parties and their associated costs, as well as the inherent risk of fraud. Nakamoto addressed these issues by presenting "an electronic payment system based on cryptographic proof" that does not require trusted third parties [5]. By articulating this solution, Nakamoto developed the underlying technical mechanisms of Blockchain technology and its first application, a cryptocurrency known as Bitcoin.

Blockchain operates using peer-to-peer networking and is designed to capture, verify, and store data in a block-based database structure. This is how Blockchain derives its name. Its structure can simply be described as a chain of blocks; all data is stored in blocks and each block is interlinked with one other block by cryptographic references. Considering that each block has the potential to store thousands of data submissions, the interconnection of blocks forms a Blockchain. It is important to recognise that not all types of Distributed Ledgers function in the same way and will often have unique methods for verifying data, storing data and structuring databases.

6.2.2   How Blockchain Works

Public Blockchain databases intentionally provide global public access to stored data for the purposes of data transparency, integrity, and auditability [3]. Consortium and Private Blockchains also exist which provide varying levels of accessibility [6]. Regardless of the type, Blockchain utilises a unique process to ensure that all broadcast data is properly verified before it is committed to a distributed database, where data records become immutable [7].

In order to commit data to the Blockchain database, the first stage involves pushing valid data onto the Blockchain network. In several applications of Blockchain, public key cryptography is applied, forcing data submissions to be privately signed; this facilitates verification of individual data submissions and their integrity using public keys [7]. Once a data submission has entered the network, it will be added to a collection of unconfirmed data submissions. Data submissions exist in this pool until a potential block can be formed, where hundreds or thousands of data submissions are assembled together. The newly assembled block will initially contain all of the collated data submissions and the hash of the current leading block in the Blockchain. The hash of the new block will then be calculated, represented as a Merkle Tree for improved processing efficiency, and added to the new block. Using this method to define the new hash, the newly assembled block will become bound to the current leading block, establishing the next potential block in the chain. Following this, the hash of the new block will then be passed to a Blockchain timestamp server to widely publish the hash and add timestamp information [5].

With the new block publicly announced across the Blockchain network, all associated nodes will attempt to verify the block using the supported consensus algorithm of the Blockchain instance, unless it is recognised as the Genesis block; this is the first block on the Blockchain network and does not require verification [7]. If successfully verified, the new block is committed to the distributed Blockchain database, otherwise the block is rejected and discarded. The Blockchain with the longest branch of blocks is acknowledged as the most authentic chain.

6.2.3   Consensus Algorithms

Consensus algorithms are vital in Blockchain technology as they provide distributed nodes with a universal mechanism for automatically verifying data submissions and determining the latest copy of the distributed ledger. By fulfilling this purpose, they establish integrity and governance for the entire Blockchain, though these are not without benefits and drawbacks.

The consensus algorithm applied often varies across Blockchain instances as each mechanism has unique characteristics and principles, making them ideal for different applications [8]. Several prominent approaches to Blockchain consensus include Proof of Work (PoW), Proof of Stake (PoS), and Delegated Proof of Stake (DPoS).

Proof of Work is one of the most commonly used consensus mechanisms on Blockchain. To validate a block, dedicated mining nodes are required to carry out complex hashing calculations that are challenging to generate, but simple for other nodes to verify. When a mining node finds the cryptographic solution, the solution is broadcast on the network for verification by other nodes. Upon success, the block is committed to the distributed Blockchain and the mining node is rewarded for their efforts [9]. Whilst this algorithm is recognised for being very secure, the amount of compute required is an area of controversy; this is an inefficiency which has financial and environmental impacts.

Proof of Stake was designed as an efficient alternative to Proof of Work, resolving both cost and energy consumption issues. Unlike Proof of Work, the principle of this consensus algorithm focuses on the ownership of currency, in that the amount of currency held by a node directly increases its probability of being selected as a block validator. This approach requires no mining capability, but it can be deemed almost plutocratic, with a firm bias towards dominant entities with large amounts of currency. Zheng et al. in [9] state that selection modifiers like randomisation and coin age have been incorporated in some instances in an attempt to mitigate this preferential system.

Delegated Proof of Stake is a variant of Proof of Stake that adopts a more demo-cratic focus, while remaining efficient. Stakeholders are granted the ability to vote and elect nodes as delegated block validators, relative to their ownership of currency [9]. In the same way, this should give stakeholders the ability to remove delegated block validators if they act maliciously, obligating block validators to uphold their reputation in order to continue validating data and receiving rewards. This consensus algorithm is thought to be quicker and handle greater amounts of data than the Proof of Work and Proof of Stake mechanisms.

6.3   Blockchain Pros and Cons

- Distributed without Intermediaries: Data is widespread across nodes with no reliance or trust in an intermediary [10]. By removing intermediaries, costs are removed, data fees are lower and data submissions are automatically verified.
- Immutable and Transparent: Data committed to the Blockchain is stored indef-initely and can be made clear for all nodes to observe. This data is designed to never change, though this is challenging but possible under majority control.
- Inherently Secure:By encompassing public key cryptography, deep integration of block signatures, validating data submissions with consensus mechanisms, and interlinking data storage, security is integral to Blockchain which ensures it is highly resistant to attacks.
- Anonymous: By implementing mechanisms like consensus, Blockchain nodes are trust less. By trusting in the principles of the technology, nodes are publicly distinguishable but cannot be linked to an identity.

- Auditable Integrity: Due to the process around the submission, verification, and storage of data, Blockchain records are highly reliable and accurate. Combined with immutability, records can be openly examined and traced between nodes.
- Limited Scalability: The balance between decentralisation, security, and scalability is recognised as the Blockchain trilemma by Ethereum founder, Vitalik Buterin. Through enforcing decentralisation and security, scalability becomes constrained, directly effecting speed. However, enhancements to scalability and speed may still be achievable using solutions that extend off the Blockchain [11].
- Storage and Energy Inefficiencies: In distributed technologies, nodes bear significant inefficiencies. As each stores a copy of the Blockchain, storage consumption is increased, which amounts greatly when aggregated. Additionally, nodes that participate in demanding validation suffer from extensive energy usage which is financially and environmentally impacting.
- Complex Developing Technology: Blockchain is still considered to be a developing technology, despite Nakamoto efforts to make it a modern reality, over a decade ago. Amongst other issues, its slow integration may be partially due its underlying complexity and a lack of expertise [12].
- Perception of Criminality: The technology has long been criticised and perceived as a facilitator for nefarious activity, likely due to its anonymity, which undoubtedly damaged its initial reputation. Blockchain's reputation is improving with an expansion of understanding and a growing appreciation of its characteristics [13].
- Lack of Regulation: Regulation has significantly affected the acceptance of Blockchain, especially as it confronts the fundamentals of economy and society. The lack of governance over the technology has hindered integration and has led many to build distrust, particularly with frequent issues involving initial coin offerings.

6.4 Applications of Blockchain

Blockchain has rapidly evolved from Bitcoin, proposed by Nakamoto in 2008, into a technology that aims to challenge all sectors of industry, potentially impacting revenue, cost, capital, and social aspects [14]. Zhao et al. in [9] introduces the idea of three generations of Blockchain: "Blockchain 1.0 for digital currency, Blockchain 2.0 for digital finance, and Blockchain 3.0 for digital society". These generations are arguably very accurate and closely represent the critical stages of Blockchain over the past decade, although Blockchain 3.0 is still emerging.

6.4.1 Blockchain 1.0

The first application of digital currency in Blockchain technology was developed in the year following Satoshi Nakamoto's 2008 whitepaper. This discussed the concept of "A Peer-to-Peer Electronic Cash System", named Bitcoin, which was developed by Nakamoto and publicly launched in 2009. As Nakamoto outlined, this was a digital cryptographic currency system, designed to operate in a decentralised manner without the need for intermediaries. Since its release, Bitcoin has grown to become the most famous Blockchain cryptocurrency, with over 5000 other cryptocurrencies launched, as of February 2020 [15]. This demonstrates the influence Bitcoin has had in the cryptocurrency market, now estimated to have a total market capitalisation of 220 trillion GBP.

6.4.2 Blockchain 2.0

Overtime, growing limitations of Bitcoin have been identified. Its limited functionality as a cryptocurrency compelled Vitalik Buterin to introduce Ethereum in 2014, outlined in the whitepaper "A Next-Generation Smart Contract and Decentralized Application Platform" [16]. Unlike Bitcoin, however, Ethereum was founded as a public decentralised platform, built on the concept of Blockchain. It is the first Blockchain to facilitate the development and deployment of smart contracts; self-executing contracts that are only triggered when precise conditions are met. These are best written using the Solidity programming language and can be used to create distributed apps, and even decentralised autonomous organisations [17, 18].

The financial industry has registered significant interest in Blockchain 2.0 as it offers serious benefits to the sector [19]. This is set to bring changes to payments, trading and account management which will inevitably reduce costs and streamline banking processes. However, many financial companies have chosen not to publicly disclose their plans or Blockchain partners, which is likely due to the high competition naturally found within this sector. Despite this, some companies have chosen to outline their strategies. Ripple has revealed that Santander is the first bank in the UK to partner with Ripple's Blockchain, whilst IBM has announced a partnership with Stellar's Blockchain. These partnerships have been established to enable faster cross-border payments, though further integration is expected [20].

6.5 Blockchain Security Issues

The rise of Blockchain technology has been largely successful; however with its existence spent openly operational, it is unsurprising that this has caused challenges. Security vulnerabilities are unavoidable and have been recognised in many aspects of Blockchain, including peer-to-peer networks, consensus mechanisms, mining, smart contracts, and wallets, many of which have led to real-world attacks. Several of the most notable vulnerabilities, with recognised examples, have been discussed.

6.5.1   Cryptojacking Attacks

Cryptojacking is type of mining attack that has become prevalent with the growth of the cryptocurrency market and the broad adoption of the proof of work consensus mechanism. By maliciously hijacking computer resource, unsuspecting systems can be instructed to mine a particular Blockchain on behalf of an attacker, rewarding cryptocurrency for each successfully mined block [21, 22]. This incentivizes the attacker to hijack a pool of systems as this would be substantially more effective and yield greater reward.

In early 2018, cryptojacking activity peaked with over 8 million cryptojacking events blocked by Symantec, as published in Symantec's 2018 Internet security threat report cryptojacking whitepaper. It is important to acknowledge that Symantec provides regular security reports. This highlights the intelligence and analysis capabilities of the company and may act as advertisement to prospective customers. The report recognised two types of cryptojacking, file-based and browser-based.

CoinHive was launched in September 2017, developed as a browser based JavaScript file with the ability to initiate mining of the Monero cryptocurrency [21]. Whilst this solution was designed as an alternative stream of revenue, as opposed to advertisements, the technology could easily be exploited by website owners and hackers due to its usability and stealth. Issues became apparent when websites were found to include mining scripts without consent, enabling all visitor devices to be targeted. The mining scripts caused decreased performance and increased energy usage and heat in devices, though this can be remediated by using anti-malware tools, capable of mining detection. File-based cryptojacking operates similarly but is installed on systems directly by malware, like Smominru, a trojan coinminer [23].

6.5.2   51% Attack

The 51% attack exists because of a technological oversight by Nakamoto, who deemed it "computationally impractical" for Blockchain to be controlled by a majority. This type of attack can only be carried out against Blockchains that apply the Proof of Work consensus algorithm; it poses significant harm to smaller Blockchain networks with reduced hashing power, as little resource is needed form a majority.

To initiate this attack, an individual or pool of mining nodes must control at least a 51% majority of the Proof of Work computing power. This means that the hash can be calculated quicker than other nodes, thus taking control of the Blockchain [24]. Once Blockchain control is achieved, it is possible for attackers to spend currency twice by manipulating or reversing transactions, otherwise known as double-spending, a common issue present in traditional finance systems.

While Bitcoin has not faced exploitation of this vulnerability, a derivative known as Bitcoin Gold has seen numerous attacks since its hard fork from Bitcoin in October 2017. In May 2018, Bitcoin Gold was struck by a series of 51% attacks, utilised to perform double spending, which stole approximately 18 million from cryptocurrency exchanges [25]. More recently in January 2020, Bitcoin Gold suffered more 51% attacks which saw a further 70000 stolen [26]. This highlights that these attacks are still possible despite the computational challenges faced from an increased Blockchain network, which may suggest that rentable mining power services, like NiceHash, could be partially to blame [27]. To combat this vulnerability, rentable mining services should be forced to take more responsibility, though 51% attacks can be mitigated entirely by adopting an alternative Blockchain consensus algorithm.

6.5.3 Smart Contract Attacks

As previously highlighted, Ethereum was the first Blockchain to support the development of smart contracts, which can be combined to form distributed apps [18]. Once implemented into the Blockchain, the contracts adopt immutability, therefore the code becomes unchangeable [28]. This provides substantial security benefits but can obstruct remediation when programming-based security incidents arise.

In early 2016, several developers conceptualised the idea of a decentralised autonomous organisation (DAO) on the Ethereum platform, known as the "The DAO" [29]. A decentralised business model, driven entirely using smart contracts, whereby investors and contributors would be granted votes towards future projects and direction. By June 2016, the project had raised over 150 million through Ether crowdfunding, from over 11000 contributors [30].

Within weeks of crowdfund completion, The DAO was attacked. This targeted a vulnerability in the underlying smart contract code which allowed the attacker to recursively withdraw Ether from the crowdfund into their own possession, estimated to total 50 million [31]. Following exhaustive discussions, the incident was controversially rectified by executing an Ethereum hard fork, restoring the network to a state prior to the attack, whilst harming the principle of immutability. This enabled all funding to be returned to its respective investors, with "The DAO" suffering catastrophic financial and reputational loss [28].

6.6 Discussion

Blockchain is a revolutionary technology that has undergone rapid evolution within the past decade, now anticipated to drive extensive transformation. It aims to distribute and decentralise data by eliminating intermediaries, requiring members of the network to trust in the technology instead, which actively seeks to protect the anonymity, privacy and security of data.

By performing this critical literature review, it is evident that the concept of Blockchain and its underlying technologies are widely understood. Being built on the principles of cryptography with data submissions verified by consensus, the security benefits are largely recognised, meaning that verified data records feature immutability, transparency, integrity, and auditability. However, the drawbacks do not seem to be as well acknowledged; it is evident that there are new inefficiencies, perceptions to overcome, and regulations to introduce, though these seem somewhat concealed in the current excitement surrounding the technology. It is also apparent that Blockchain has some prominent security vulnerabilities that can have severe impacts on the technology, finance, and society, but mitigation and remediation plans exist to reduce their effects or greatly increase security complexity. Considering both sides of the discussion, the benefits of Blockchain seem to substantially outweigh the drawbacks and issues, with aims to positively influence a wide array of sectors and industries like agriculture [32], drones [33], energy [34], industrial networks [35] or even voting process [36].

6.6.1 Voting and Blockchain

In the article in [36], the authors proposed a novel blockchain-based voting mechanism that consists of three major components:

1. **Smart Contracts** As we mentioned in a previous section, contracts are a piece of program which run on top of blockchain. They can be sued in order to remove any participation from third party into the entire process. In this specific scenario, smart contracts can be used in registration of the voters.
2. **Voter Registration and Verification**: In cryptocurrency system, any individual that becomes part of the network is assigned a wallet wherein he/she keep their network specified cryptocurrencies. This technology can be used for initial step of Voter Registration. Voter Verification is also important aspect of the process. Just registration of the voter will not alone provide a foolproof system. When the voter is casting the vote, appropriate measure needs to be taken to ensure the voter who is voting is indeed him/her and not someone else who is voting on their behalf.
3. **Voting System flow and Voting Machinery**: These include any computer, mobile phone, or government-approved devices.

6.6.2 Energy and Blockchain

Using blockchain technology, plants where electricity is produced are linked to specific points of consumption, allowing the source of the energy to be traced and ultimately encouraging the use of renewable energy. The transparency of the system guarantees, in real time, that the energy supplied and consumed is 100% renewable. People can sell their produced electricity with each other without wasting it, creating a small economy without middleman. As proposed in [34], a blockchain-based energy exchange system that incorporates a reliable peer-to-peer energy system can be deployed in the near future. In order to prevent smart grid attacks, the proposed framework makes the generation of blocks using short signatures and hash functions and also incorporates traditional security mechanisms like an IDS.

6.6.3   Web 3.0 and Blockchain

The core structure and architecture of the backend of the web will change dramatically over the next years. Blockchain technology is said to be a key player that is going to play a major role in the transformation of centralised Web 2.0 to a decentralised Web 3.0. For instance, in social media people will no longer connect and interact with each other through huge centralised servers of corporations, but instead there will be multiple blockchains running the applications. The control over the data will be stored using private keys on the blockchain.

6.7   Conclusions

Despite the advancements made, Blockchain should be considered a developing technology, implied by the continued need for research and development and, at this time, limited public implementation. However, this is not always apparent due to the significant hype that seems to shroud the technology and the cryptocurrency market. Enterprises appear to have recognised Blockchain's immaturity which has undoubtedly hindered its adoption and real-world use cases. For this reason, it is possible that enterprises may be exploring and experimenting with their own private implementations of Blockchain, where they have greater control, before publicly disclosing their strategies. It is important to recognise that the distributed approach offered by Blockchain will not be suitable for all applications; therefore, centralised solutions will continue to be applicable in many use cases.

In the short-term, we should expect several key improvements. Blockchain 3.0 aims to develop resolutions and workarounds for current scalability issues, which will involve extending off the blockchain, as seen in emerging solutions like Plasma and the Lightning Network (Coinspace, 2019). In addition to this, we should expect Blockchain to be applied to more use cases which will coincide with the introduction of more regulation and governance.

Longer term, adoption of the technology should be even more prevalent, with widely documented cases, directly challenging current configurations. This could eventually cause a shift towards a new societal norm, with reduced intermediaries and increased security. However, the dawn of quantum computing will inevitably threaten existing cryptography, obligating Blockchain to swiftly adapt.

References

1. Belin, O.: The difference between blockchain and distributed ledger technology (2008)
2. Kehrli, J.: Blockchain explained. Netguardians [en línia].[Data de consulta: 25 de juny de 2017]. https://www.netguardians.ch/news/2016/11/17/blockchain-explained-part-1 (2016)
3. Ferrag, M.A., Maglaras, L., Janicke, H.: Blockchain and its role in the internet of things. In: Strategic Innovative Marketing and Tourism, pp. 1029–1038. Springer, Berlin (2019)
4. Natarajan, H., Krause, S., Gradstein, H.: Distributed Ledger Technology and Blockchain. World Bank (2017)
5. Nakamoto, S.: Bitcoin: A peer-to-peer electronic cash system. Technical report Manubot (2019)
6. Lin, I.-C., Liao, T.-C.: A survey of blockchain security issues and challenges. IJ Netw. Secur. **19**(5), 653–659 (2017)
7. Drescher, D.: Blockchain Basics, vol. 276. Springer, Berlin (2017)
8. Mingxiao, D., Xiaofeng, M., Zhe, Z., Xiangwei, W., Qijun, C.: A review on consensus algorithm of blockchain. In: 2017 IEEE International Conference on Systems, Man, and Cybernetics (SMC), pp. 2567–2572. IEEE, New York (2017)
9. Zheng, Z., Xie, S., Dai, H., Chen, X., Wang, H.: An overview of blockchain technology: Architecture, consensus, and future trends. In: 2017 IEEE International Congress on Big Data (BigData Congress), pp. 557–564. IEEE, New York (2017)
10. Golosova, J., Romanovs, A.: The advantages and disadvantages of the blockchain technology. In: 2018 IEEE 6th Workshop on Advances in Information. Electronic and Electrical Engineering (AIEEE), pp. 1–6. IEEE, New York (2018)
11. Im, D.K.D.: The blockchain trilemma (2018)
12. Chen, C., Wang, C., Hou, J., Wang, L., Zhang, P., Xu, R., Qu, X., Zhang, H.: Strategy of training blockchain talents in application-oriented universities: A case study (2019)
13. Khan, M.A., Salah, K.: IoT security: Review, blockchain solutions, and open challenges. Future Gener. Comput. Syst. **82**, 395–411 (2018)
14. Carson, B., Romanelli, G., Walsh, P., Zhumaev, A.: Blockchain beyond the hype: What is the strategic business value. McKinsey & Company, pp. 1–13 (2018)
15. Zheng, Z., Xie, S., Dai, H.-N., Chen, X., Wang, H.: Blockchain challenges and opportunities: A survey. Int. J. Web Grid Services **14**(4), 352–375 (2018)
16. Buterin, V., et al.: A next-generation smart contract and decentralized application platform. White paper **3**(37) (2014)
17. Kolb, J., AbdelBaky, M., Katz, R.H., Culler, D.E.: Core concepts, challenges, and future directions in blockchain: A centralized tutorial. ACM Comput. Surv. (CSUR) **53**(1), 1–39 (2020)
18. Vujicic, D., Jagodic, D., Randic, S.: Blockchain technology, bitcoin, and ethereum: A brief overview. In: 2018 17th International Symposium Infoteh-Jahorina (infoteh), pp. 1–6. IEEE, New York (2018)
19. Guo, Y., Liang, C.: Blockchain application and outlook in the banking industry. Finan. Innov. **2**(1), 24 (2016)
20. Long, M.: Santander becomes the first UK bank to use ripple for cross-border payments. Ripple blog (2016)
21. Musch, M., Wressnegger, C., Johns, M., Rieck, K.: Thieves in the browser: Web-based cryptojacking in the wild. In: Proceedings of the 14th International Conference on Availability, Reliability and Security, pp. 1–10 (2019)
22. Tuttle, H.: Cryptojacking. Risk Manage. **65**(7), 22–27 (2018)
23. Sigler, K.: Crypto-jacking: how cyber-criminals are exploiting the crypto-currency boom. Comput. Fraud Secur. **2018**(9), 12–14 (2018)
24. Xu, J.J.: Are blockchains immune to all malicious attacks? Finan. Innov. **2**(1), 1–9 (2016)
25. Analytica, O.: Blockchain's cybersecurity needs greater attention. Emerald Expert Briefings, no. oxan-db
26. Moroz, D.J., Aronoff, D.J., Narula, N., Parkes, D.C.: Double-spend counterattacks: Threat of retaliation in proof-of-work systems. arXiv preprint arXiv:2002.10736 (2020)

27. Yuen, M.-C., Lau, K.-M., Ng, K.-F.: An automated solution for improving the efficiency of cryptocurrency mining. In: 2020 International Conference on COMmunication Systems & NETworkS (COMSNETS), pp. 1–5. IEEE, New York (2020)
28. Sayeed, S., Marco-Gisbert, H., Caira, T.: Smart contract: Attacks and protections. IEEE Access **8**, 24416–24427 (2020)
29. Mehar, M.I., Shier, C.L., Giambattista, A., Gong, E., Fletcher, G., Sanayhie, R., Kim, H.M., Laskowski, M.: Understanding a revolutionary and flawed grand experiment in blockchain: the Dao attack. J. Cases Inform. Technol. (JCIT) **21**(1), 19–32 (2019)
30. Siegel, D.: Understanding the Dao attack. Retrieved June, vol. 13, p. 2018 (2016)
31. Atzei, N., Bartoletti, M., Cimoli, T.: A survey of attacks on ethereum smart contracts (SOK). In: International Conference on Principles of Security and Trust, pp. 164–186. Springer, Berlin (2017)
32. Ferrag, M.A., Shu, L., Yang, X., Derhab, A., Maglaras, L.: Security and privacy for green IoT-based agriculture: Review, blockchain solutions, and challenges. IEEE Access **8**, 32031–32053 (2020)
33. Ferrag, M.A., Maglaras, L.: Deliverycoin: An ids and blockchain-based delivery framework for drone-delivered services. Computers **8**(3), 58 (2019)
34. Ferrag, M.A., Maglaras, L.: Deepcoin: A novel deep learning and blockchain-based energy exchange framework for smart grids. IEEE Trans. Eng. Manage. (2019)
35. Derhab, A., Guerroumi, M., Gumaei, A., Maglaras, L., Ferrag, M.A., Mukherjee, M., Khan, F.A.: Blockchain and random subspace learning-based ids for SDN-enabled industrial IoT security. Sensors **19**(14), 3119 (2019)
36. Soni, Y., Maglaras, L., Ferrag, M.A.: Blockchain based voting systems. In: ECCWS 2020 19th European Conference on Cyber Warfare and Security. Academic Conferences and publishing limited (2020)

Chapter 7
Personal Data Protection in Blockchain with Zero-Knowledge Proof

Seval Capraz and Adnan Ozsoy

Abstract Blockchain is a trending research topic and provides a high level of security, anonymity, and data integrity without a trust. One of the usage areas of blockchain is personal data protection. There are solutions proposed in literature for this specific need because personal data protection is still a problem in this big data era. It is obvious that the more personal and sensitive data are collected, the harder the problem to keep them secure, stable, and under control. Blockchain comes with significant advantages like verified and trackable transactions which allow anyone to follow the asset trivially. In addition to this, blockchain can be a tool to keep data itself anonymously. The researchers take advantages from blockchain to create more secure personal data protection systems. In this study, we are looking for the best personal data protection system with blockchain solutions. So we summarize proposed models and realize that a model based on off-blockchain with zero-knowledge proof is the best choice. Nowadays, zero-knowledge proof method's usage on blockchain is popular. We explain in this study why it is the best choice based on personal data protection and how we can benefit from it. The goal of this study is to obtain a general view of the characteristics and usefulness of blockchain for more secure personal data protection reported in scientific literature and used in industry.

7.1 Introduction

Personal data protection is an area that interests everyone. All the systems we use everyday collect, use, and keep our data. Can we say that all types of data are personal? We can classify data as personal data if it relates to a living individual. For example, not only a password but also name and surname, a telephone number, an email address or an IP address are all personal data. The systems which collect

S. Capraz (✉) · A. Ozsoy
Department of Computer Engineering, Hacettepe University, Ankara, Turkey
e-mail: seval.capraz@hacettepe.edu.tr

A. Ozsoy
e-mail: adnan.ozsoy@hacettepe.edu.tr

© The Author(s), under exclusive license to Springer Nature Singapore Pte Ltd. 2021   109
S. Patnaik et al. (eds.), *Blockchain Technology and Innovations in Business Processes*,
Smart Innovation, Systems and Technologies 219,
https://doi.org/10.1007/978-981-33-6470-7_7

these data should have a proper personal data protection system that provides a few important principles. These principles are mostly defined in law in some countries or states. For instance GDPR in Europe, KVKK in Turkey, CCPA in California etc. We determined the key principles of personal data protection based on literature. We can summarize the principles in three main topics. First of all, the system should provide data ownership. Personal data belongs to owners. Only owners have the right to control the data. The other sources who want to reach the data need permission from the owner. Secondly, individuals have the right to learn their collected data. Most of the systems we use daily collect information about us without our knowledge. They collect the data in the background quietly and they store the data in their storages. So owners should learn anytime their personal data records are being collected. In addition to this, we mostly do not know which applications, systems and services access our data. The owners should learn anytime how their data records are accessed. This means data transparency should be provided between systems and owners. Thirdly, only the owners have the ability to change permissions on their personal data. In a proper personal data protection system, individuals can change the set of permissions. They can give or revoke access to previously collected data. These three concerns should be provided in proposed systems no matter what technology is used.

Personal data protection systems can be modeled and implemented in a lot of ways. The earlier solutions for personal and sensitive data protection include five methods [1]. First and the most desired feature of them is anonymization. It is called k-anonymity. If anonymization techniques are applied on personal data, the data alone can not relate to any individual. So it is no longer considered as personal data. This is very important in case data records are stolen. When attackers access a personal data, they should not find the owner's real identity. If they find the owner's identity, they can use it for their bad intentions. For example, they can use the data to access the owner's bank account. Anonymization is an important key feature and it has been supported for many years. In a proper system, the anonymisation must be irreversible. For example storing hash functions for passwords on a database hides data itself and nobody can generate the original password from the hash of it. The second method is using a diverse enough set of possible values to represent data. It is an extension to k-anonymity. It includes l-diversity. If a system generates possible values and stores all of them, the owners identity cannot be detected. The third method is distributing sensitive data. It is called t-closeness. It uses distributed systems. The system parses the original data into many pieces and stores every part of it in different locations. An attacker can not obtain meaningful data unless he or she has whole parts of original data. The fourth method is adding noise to the data. It is known as differential privacy. The noise makes data unrecognizable. The last method is homomorphic encryption which is a technique used to operate on encrypted data without decrypting it. This method uses heavy mathematical calculations, so it is slower than other solutions.

These earlier solutions are not practical and reliable. There are mainly four reasons why earlier solutions are not useful and secure enough. Firstly de-anonymization techniques can be used to reach personal data. Unauthorized people can reach data thanks to the de-anonymization techniques, so it is not a desired result. Secondly, all of the earlier solutions are bound to be a centralized system. In a centralized system, users cannot fully control their personal data. Some centralized systems do not allow owners to control their personal data. We need a distributed system to give full control of data to data owners. Thirdly, there are security issues of such systems and difficulties of implementation, especially in homomorphic encryption. Fourthly, there should be a third-party system to keep users' personal data and people have to trust it. It is very difficult to keep personal data safe in this kind of third-party system. They can use the data for any aim without the owner's permission. The privacy issues are addressed in the following sections in detail.

In recent years, new solutions emerged based on distributed ledger technology (DLT). A DLT is a digital data which is spread across the world. It is based on consensus. A new entry can be included in the ledger if all participants agree. The ledgers are immutable and there is no central administrator. This new technique removes the dependency on third-party systems. For instance, blockchain is a kind of DLT and used mainly in the finance sector. Banks are the third-party trust systems for money transactions; however, we do not need banks to send money from an individual to another with blockchain. It is a more secure system than traditional banking since blockchain is decentralized and the history is not changeable. These features are wanted in a secure personal data system as well. Therefore, we can use blockchain in a personal data protection system.

In a proper personal data protection system, we need the highest security. There-fore, we searched for studies about blockchain solutions for personal data protection reported recently. We summarized the studies and analyzed the trend. The trend is using blockchain as a tool to handle history and permissions. There are studies proposed in scientific literature about the benefits of blockchain on personal data protection topics. Among proposed methods, the most secure blockchain systems use zero-knowledge proof. It is a known technique to prove you have the information by hiding the original information. Zero-knowledge proof impairs the overall perfor-mance of the system however it provides the highest security. Most of the studies use off-blockchain type to store data. **In this chapter, we study the state-of-the-art personal data protection systems that use blockchain technology, summa-rize proposed models based on off-blockchain with zero-knowledge proof**. This study aims to help researchers and practitioners to understand how to use blockchain for personal data protection and explain the usage of zero-knowledge proof for the most secure method which is easy to implement and serve different purposes. This article primarily includes the following sections: related works, background about blockchain, personal data protection and blockchain, zero-proof knowledge on blockchain and conclusion.

7.2 Related Works

Six years later than the invention of blockchain by Nakamoto [2], the first study which was proposed by Zyskind et al. [1] about the usage of blockchain in personal data protection systems emerged. In this first study, they combine blockchain and off-blockchain storages and construct a personal data management platform focused on privacy. The proposed system focuses on mobile phone users and services. Mobile applications have access to user information that is personal, so blockchain is used as a tool to manage personal data protection. Following this study, many other studies were published as we describe them in detail in the following paragraphs.

In the proposed system of Zyskind et al. [1], there are two types of users: users and applications. Users are the owners of personal data and services want to access data with permission. The data consist of user data like location, contacts, etc. There is a blockchain which runs data policies. The data is stored in a off-blockchain key-value store (such as by a distributed hash table) and the references are hold on blockchain. The blockchain holds the SHA-256 hash of data as a pointer. The data can be hold in any kind of system like central, distributed, local, or cloud. That does not affect the blockchain. The proposed system has pseudo identities with public key and private key. It has also combined identities like multiple guests with a single owner. Blockchain works with two different types of transactions: *Taccess* and *Tdata*. For access management, a new data policy can be sent to blockchain with a structure of {User, Application, Data Type} which is *Taccess*. This command can change the permissions of data access, or replaces permissions when you send again. A new data can be stored and retrieved with *Tdata* command. To store a data, personal user data should be encrypted before sent. Hash information of the data are written on the blockchain. The original data is stored in a repository outside of the blockchain. Then, the data can be queried with the corresponding keys.

In 2017, Liang et al. [3] proposed ProvChain which is a blockchain-based cloud data provenance system. Thanks to the blockchain, it provides tamper-proof records and transparency. It is different from other solutions with usage of clouds. The same author Liang et al. [4] proposed a new system next year. It is a new cyber-physical system using blockchain. In this paper, it is indicated that blockchain is capable of assured data provenance validation and user privacy preservation.

Kuner et al. [5] discussed blockchain versus data protection in their study. They give the disadvantages and advantages of blockchain based on European Union's general data protection regulation (GDPR). The disadvantages are the same with others. The main concern is blockchain and is not compatible with data protection. However, this concern is irrelevant. Kuner's suggestion includes three main advantages. First, blockchain can be set up by a consortium that is governed by rules determining which party processes any data. By this way, privacy is provided. Second, consensus by authority can be defined, so the new blocks have to get the permission of this authority to join the chain. Third, the ledgers can be editable if it is defined in this way.

In medical world or health sector, security and how their patients' personal data is stored are very important. There are many proposed studies which describe blockchain for health data and its potential use in health IT and healthcare-related research [6–9] in different years. Mainly they proposed a blockchain-based security and access control manager for health records. Access control is a well-known problem in the health sector. Blockchain provides more security for health IT. They did not focus on privacy; therefore, those proposed methods can be improved many ways.

In 2017, Ramachandran and Kantarcioglu [10] proposed using blockchain and smart contracts for secure data provenance management. Next year, they implemented and proposed SmartProvenance system [11] as a distributed and blockchain-based data provenance system. In the same year, Joshi et al. [12] published a survey on security and privacy issues of blockchain technology.

In 2018, Vishwa and Hussain [13] proposed a new approach with blockchain for the multimedia sector. They used blockchain for privacy concerns and provenance. That study seems to use the similar method of [1] like using off-blockchain for keeping history of transactions in order to access to a node and data retrieval. They also gave a reference of [1] which they used as their method as a data protection system for multimedia data. In 2018, Jia et al. [14] used blockchain in crowd sensing networks. Their aim is to create a tamper-proof system in location privacy protection incentive mechanism.

In 2019, Feng et al. [15] published a survey which inspects privacy protection in blockchain systems. They divided privacy requirements for blockchain into two groups: identity privacy and transaction privacy. We influenced that study very much. The comparison of methodologies based on privacy protection in the blockchain system is given in Fig. 7.1. They concluded their study by finding the best solution of privacy for both groups: using zero-knowledge proof on blockchain. This method hides transaction relationship and content. The one and only disadvantage of it is heavy computation overhead. There are two popular existing projects which use zero proof knowledge in blockchain: Zcoin [16] and Zcash [17]. We also summarized their protocols lying under them in our study: Zerocoin and Zerocash. That survey indicated that the future directions for blockchain will tend to use more zero-knowledge proof approaches on blockchain. So we decided to examine this approach more.

In cryptocurrencies, in order to obscure the trail of cryptocurrency transactions, third-party coin mixing services can be used. Miers et al. [18] proposed the Zerocoin protocol in 2013. It allows anonymization of cryptocurrency transactions without going through a trusted third-party. In this protocol, the history of a coin is deleted by

Methodologies	Security Objectives	Main Disadvantages	Existing Projects
Centralized mixing in Section 4.1.1	Obfuscating transaction relationship	1. Waiting delay 2. Single failure 3. Service fees 4. No protected on transaction content	Mixing Websites, Dash, etc
Decentralized mixing in Section 4.1.2	Decentralized Obfuscating transaction relationship	1. Waiting delay 2. Sybil attack 3. Heavy communication overhead 4. No protected on transaction content	CoinJoin, CoinShuffle, etc
Ring signature in Section 4.2	Hiding tradition origin	1. Limited size of anonymity set 2. Heavy storage overhead 3. No protected on data to be signed 4. No protected on transaction target	$\perp$
CryptoNote in Section 4.2.2	Hiding tradition relationship and content	1. Limited size of anonymity set 2. Heavy storage overhead	Monero
NIZK in Section 4.3 and Section 5.1	Hidding transaction relationship and content	Heavy computation overhead	Zcoin, Zcash, etc
Commitment in Section 4.3 and Section 5.2.1	Non-repudiation on protected transaction content	Need additional ways for anonymity	Confidential Transaction (CT)
HC in Section 5.2.2	Hidding transaction contents	No support for auditing	$\perp$

Fig. 7.1 Comparison of methodologies based on privacy protection in the blockchain system by Feng et al. [15]

destroying the coin. Then the system mints again a new coin with no history behind. Nobody can find which coin is being spent because its original history is deleted. Next year after Zerocoin, with similar but not the same the approach, Zerocash, was proposed by Sasson et al. [19] in 2014. Their cryptographic algorithms are different but both use zero-knowledge proof.

One of proposed systems which use zero-knowledge proof for personal data protection on blockchain is Hawk [20]. Hawk is a smart contract system where programmers can use. Unlike similar systems, Hawk provides transactional privacy by not storing financial transactions on the blockchain. Programmers do not need to use cryptography for their smart contracts, the system automatically does for them. The information on blockchain is encrypted and can be validated by zero-knowledge proofs. New private coins are sent to the blockchain along with zero-knowledge proofs. The system uses the approach of Zerocash [19]. Hawk uses a modified version of Zerocash and uses a coin-based blockchain.

7.3 Background About Blockchain

Blockchain technology is a public ledger, a distributed software infrastructure that provides a decentralized database. It constitutes a chain of blocks where we record and distribute the digital information. The blockchain keeps information of transactions on blocks that cannot be edited anymore. Blockchain consists of multiple blocks followed by each other like a chain. There is no need for a trust in blockchain to join or use the system. Anyone can join the chain and share information without any need for a central controlling system. Bitcoin is the first type of blockchain-based system [2]. When a block stores new data and when it is validated in a consensus algorithm, it is added to the blockchain. Blockchain has four architectural components: record, block, P2P network, and consensus algorithm.

First of all, blockchain consists of records and blocks. Records can be any kind of data like money transactions, inventory inputs, and customer records. All new records are announced to the system as a request. When a new request is done, the miners create a new block by using the records and publish the new block to the system. The records are kept in blocks as information. All of the nodes on the system hold the blockchain and they can verify new blocks. After verification, a new block can be accepted.

Blocks are the main part of the blockchain system. They are linked together to form a chain. All blocks hold a reference of its previous block. This reference is the hash of the previous block. Records are joined together and saved in blocks by miners. For this purpose, digital signatures and cryptographic algorithms are used. If a block is changed, it can be detected easily because any change in a block causes change to the hash of it. All blocks hold a hash of previous block. In this way, nodes do not accept if both hashes are not the same with each other.

As an architectural communication model, blockchain uses a peer-to-peer distributed system (P2P) for decentralisation. P2P is stronger than a centralized network because it does not have a single point of attack or failure. Blockchain can be implemented as permission-based or permissionless for joining the network. After joining the network, all nodes have their own local copy of the blockchain. This means the whole chain is stored at many places. This method protects the blockchain from any data loss. In addition to this, users can be anonymous in such a system that they can use their private and public keys. The communication can be done via using private keys and public keys with cryptographic functions.

We need to be sure that every node holds the same chain of blocks so that they are consistent. In an ideal system, the history is not changed among nodes, they have to have the same local copy of blockchain. This means there should not be a synchronization problem. We need a consensus algorithm for the synchronization of the single ledger. Some of the consensus algorithms are proof of work, proof of stake, simplified byzantine fault tolerance, etc. In the proof of work system, computers solve a complex computational math problem to prove that they have done work. The first computer which finds the block and publishes it to the blockchain network gains a reward. Computers have to solve this hard problem to gain a reward. This process is called mining and that computer is called a miner. Miners have to spend significant amounts of power and energy to solve a math problem. This idea is found to discourage attackers from an attack. Attackers can attack the system if and only if they have the computing power of more than 50% of all computing power of the blockchain. This attack is called 51% attack. For bitcoin, the blockchain's computing power is very huge and it is not possible to overwhelm all other participants in the network. However, in little blockchains, this may be possible. So little blockchains can use a trust system to join the network. It is better to be private than public in this case.

The consensus algorithm also works to prevent from forking in the blockchain. Sometimes, two blocks are accepted at the same time by different nodes. In this situation, the consensus algorithm forces it to continue with the longest chain available. When the number of users increases in the system, the new blocks can be generated more rapidly and join the chain easily.

To sum up, blockchain is different than other data stores thanks to its properties as listed below:

- Individuals use pseudonyms not their real identities in the blockchain system. This allows anonymization.
- It is democratic. It is based on consensus. There is no third-party trust. Anybody can join the blockchain. There is no discrimination.
- The history cannot be changed in blockchain. It is immutable.
- The data or history can not be censored or hidden in blockchain. It is completely transparent.
- It is fully distributed, so it is not managed by a central system. Everybody can track whole records and verify anything. This system prevents changes in history.

There is one more feature of blockchain. We can create smart contacts on blockchain to facilitate, verify, or negotiate a contract agreement. Smart contracts of blockchain are functions that can be called by anywhere and not in a central location. They define the conditions of a set of rules. They can be useful to guarantee a greater degree of security and agreement. They regulate future transactions. Smart contracts are fully autonomous code that cannot be retracted once deployed. They can do things such as reads, writes, computations, send messages to other contracts and store data. Some examples are Proof of Existence, creation of a token, secure multi-party fund, voting mechanisms. We can use smart contracts to define a set of rules for personal data protection.

7.4 Personal Data Protection and Blockchain

In introduction, we summarized the main principles of data protection systems. We can use blockchain for three main purposes in the subject of personal data protection:

1. Data Ownership
   Owners are proprietors of data. They control the data. Services can access the data as guests. These access rights are given by owners and taken again by owners.
2. Data Transparency and Auditing
   Which data is collected and how to access the data should be completely transparent.
3. Detailed Access Control
   Applications have endless permissions which are taken at first installation such as mobile applications. Users should change set of permissions or cancel access for previously collected data.

We have to provide these three principles. Furthermore, to make the personal data protection sustainable, data lifecycle must be managed from start to finish. Also each activity must be maintained with a structure that cannot be changed backwards. Using the blockchain for personal data protection is not a challenge, but a catalyst to complete digital transformation in this area. There are a few reasons why blockchain is a useful tool for personal data protection.

Firstly, under the law on the protection of personal data, we can benefit from blockchain in different perspectives. From the data owner perspective, access management and monitoring is possible with blockchain which ensures that nothing can happen to this data without the consent of the individual subject. Moreover, it can be useful to establish accountability. Blockchain provides an opportunity to withdraw open consent of owners. From the data controller perspective, by smart contracts, blockchain can help for obligation to provide information and the purpose of processing the data.

Secondly, blockchain gives individuals the right to delete their personal data. Personal data can be held either inside of blockchain or outside of blockchain (off-blockchain) (Fig. 7.2). One of off-blockchain solutions was proposed in [1]. They did not use zero-knowledge proof on blockchain, it can be an extension to [1] so that the access information is hidden from the public. This will make it difficult to

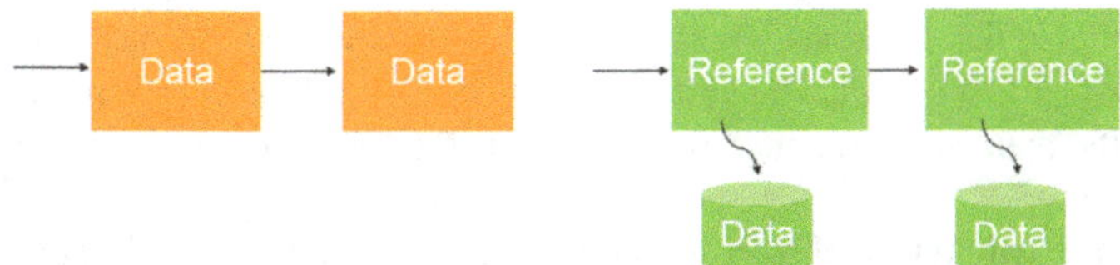

Fig. 7.2 Holding data inside of blockchain (on the left) or outside of blockchain (on the right)

find who is who. For personal data protection, it should be impossible to find who gives access rights to his or her personal data and the hash address of his or her personal data. However a verifier, let us say a service, should verify the access right of a specific person.

Because of the fact that distributed ledgers cannot be changed, deletion is technically impossible in blockchain. However, there are regulations and protocols that allow us to delete personal data from blockchain. For instance, smart contracts may include mechanisms that regulate access rights. Blockchain can be used to cancel all access rights to data. Data becomes invisible to others, even if the content is not deleted. Therefore, blockchain is beneficial to manage access rights of personal data protection.

Consequently, blockchain is flexible to solve different problems of personal data protection. We can benefit from blockchain for data ownership, data transparency and auditing and detailed access control.

7.5 Zero-Knowledge Proof on Blockchain

Sometimes, we do not want to reveal the information but we want to prove we have possession of it. There is a method we can prove that any information we have without revealing it: zero-knowledge proof (ZKP). It is a cryptographic method that was proposed in 1989 by Goldwasser et al. [21]. There are two people in this method: the prover and the verifier. The prover has possession of some information and the verifier does not want to learn the information itself but wants to know whether the prover has it or not. ZKP works very well in this problem.

For example, there is a door that works with a password and the prover knows the password. The verifier does not know the password and wants to know if the prover has it or not. The verifier wants the prover to open that door so that he or she can verify that the prover knows the password. How can we do it? We can put the door into the middle of a tunnel which has only one entrance and one exit. The door stands in the middle of the tunnel. Everybody has to open this door with its password to pass the tunnel. There will be no other way to exit from this tunnel. If the prover goes to the entrance and comes out from exit, and this means he proved that he knew the password of this door. This method is ZKP. It is found and explained mathematically in 1989 by Goldwasser et al. [21]. The other simple example of ZKP is storing hash of passwords in the database. Users can prove that they know the password entering their password as an input. The system compares the hash of input with information stored in the database. This means the verifier does not know the password but the hash of it.

There are two types of zero-knowledge proof. First one is interactive zero-knowledge proof. It is the first invention and requires multiple messages between prover and verifier. The second one is non-interactive zero-knowledge proof. It requires less interaction between the prover and verifier.

In Fig. 7.3, we described how zero-knowledge proof works [22]. It is a ZKP with a trusted third-party. The verifier verifies the prover's transaction by using a zero-knowledge proof verification algorithm. This algorithm uses the response of the prover and the proof that is generated by ZK proof construction by the prover. If the ZK Proof Verification algorithm gives a positive result, the verifier trusts the response as if it has been produced by a trusted third-party.

Blockchains provide security but not privacy. The typical type of blockchains are transparent and not totally anonymized. ZKP is required for blockchains to provide privacy. Currently, the algorithms proposed on zero-knowledge blockchain networks

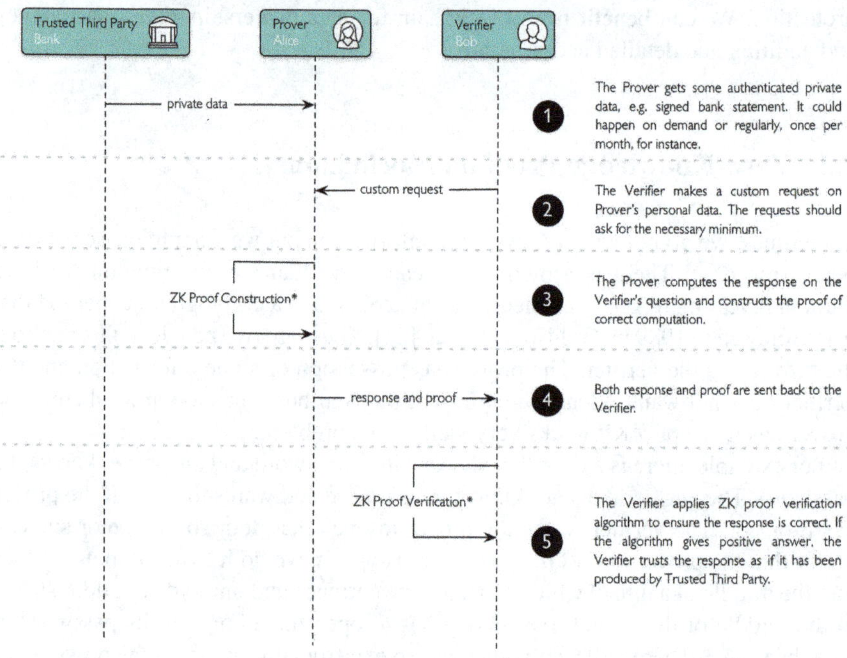

Fig. 7.3 Zero-knowledge proof on blockchain explained in summary. Figure is taken from [22]

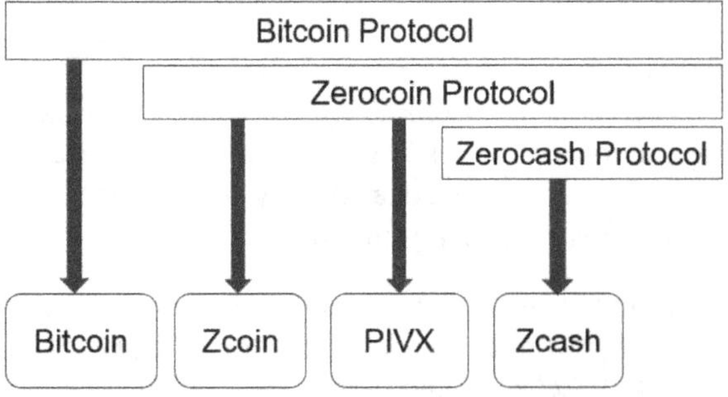

Fig. 7.4 Layered representation of Bitcoin, Zcoin, PIVX and Zcash

are very complicated and they are impractical to use on Hyperledger implementations. As opposed to this, we figured out some practical zero-knowledge algorithms that are simpler and practical to use on Hyperledger and they solve many problems of corporate and government. There are a few examples of protocols which use ZKP in blockchain. Two of them are very popular and from the electronic currency sector: Zerocoin and Zerocash.

Bitcoin is a Bitcoin Protocol-based digital currency and proposed in 2008 by Nakamoto [2]. Zerocoin Protocol is derived from Bitcoin Protocol and proposed in 2013 by Miers et al. [18]. Zcoin and PIVX currencies are based on Zerocoin Protocol. As a new layer, Zerocash Protocol is derived from Zerocoin Protocol and proposed in 2014 by Sasson et al. [19]. Zcash currency is based on Zerocash protocol. The layered representation of Bitcoin, Zcoin, PIVX and Zcash is given in Fig. 7.4. These two protocols, Zerocoin and Zerocash, rely on different cryptography, so they are totally different. These protocols are designed with four probabilistic polynomial-time (PPT) algorithms to implement ZKP: (setup, mint, spend, verify).

There are also improvements on Zerocoin with usage of 1-out-of-Many Proofs (Sigma) by Groth and Kohlweiss [23]. In [23], a new Zerocoin design is proposed. With the new design, it provides smaller proof sizes and faster verification.

Both zero-knowledge proof-based protocols have four main algorithms which is given below (algorithms are taken from Lelantus protocol [24]):

1. **Setup**: It means we need to generate a set of global public parameters. A new version of algorithms does not have a setup, so it also makes the algorithm faster.
2. **Mint**: The new block with real transaction is deleted and generated a new block with no history. This process is called minting. Coin's serial number sn is used to calculate the minted coin C. Serial number sn is needed to prevent double-spending of the coin. C is published to the blockchain and is added to the list of all previously minted coins $\{C_0, C_1, \ldots C_{n-1}\}$.
3. **Spend**: All previously minted coins $\{C_0, C_1, \ldots C_{n-1}\}$ are parsed. The proof transcript and the coin's serial number sn are published to the blockchain.
4. **Verify**: The network participants can check the validity of the block without actual inputs.

7.5.1 Zerocoin Protocol

The Zerocoin protocol is designed by using RSA accumulators and double-discrete-logarithm proofs of knowledge, which have a size exceeding 20 KB. The existence of a zero-knowledge proof system provides privacy. The algorithms are below (algorithms are taken from [18]):

- **Setup**$(1^\lambda) \rightarrow params$. On input a security parameter, run AccumSetup(1^λ) to obtain the values (N, u). Next generate primes p, q such that $p = 2^w q + 1$ for $w \geq 1$. Select random generators g, h such that $G =< g >=< h >$ and G is a subgroup of Z_q^*. Output $params = (N, u, p, q, g, h)$.
- **Mint**$(params) \rightarrow (c, skc)$. Select $S, r \leftarrow Z_q$ and compute $c \leftarrow g^S h^r \bmod p$ such that $\{c \text{ prime} \mid c \in [A, B]\}$. Set $skc = (S, r)$ and output (c, skc).
- **Spend**$(params, c, skc, R, C) \rightarrow (\pi, S)$. If $c \notin C$ output $\perp$. Compute $A \leftarrow$ Accumulate$((N, u), C)$ and $\omega \leftarrow$ GenWitness$((N, u), c, C)$. Output (π, S) where π comprises the following signature of knowledge:

$$\pi = \text{ZKSoK}[R]\{(c, w, r) : \text{AccVerify}((N, u), A, c, w) = 1 \wedge c = g^S h^r\}$$

- **Verify**$(params, \pi, S, R, C) \rightarrow \{0, 1\}$. Given a proof π, a serial number S, and a set of coins C, first compute $A \leftarrow$ Accumulate$((N, u), C)$. Next verify that π is the aforementioned signature of knowledge on R using the known public values. If the proof verifies successfully, output 1; otherwise, output 0.

For generating the parameters, Zerocoin Protocol assumes a trusted setup process. The accumulator trapdoor (p, q) is destroyed immediately after the parameters are generated because it is not used subsequent to the Setup procedure.

There are disadvantages of Zerocoin. First of all, computation time required by the process of bitcoin miners is huge. The proofs have very big sizes and keeping them on blockchain dramatically increases the size of the system. However, it can be a good practice to store proofs outside of the blockchain to diminish the size of the blockchain [18].

If a user created such a precise amount of Zerocoins in the past and no other Zerocoins with the same denomination are currently minted but unspent, this user can be detected. This harms privacy. This problem is solved by the idea of fixed denominations in the Zerocoin protocol. It works through the concepts of accumulators and zero-knowledge proofs.

Zerocoin relies on the hardness of factoring a parameter which is generated by two very large prime numbers. We can generate these values by trusted parties. We can also omit trusted parties by using RSA unfactorable objects.

Zcoin is based on Zerocoin and proposed in 2018 [16]. It replaces the proof of work system with a memory intensive Merkle tree proof algorithm. This method ensures more equitable mining among ordinary users. In the same year, a vulnerability occurred in Zcoin which allowed attackers to destroy, create or steal the coins. Then, the Lelantus protocol [24] was proposed by the Zcoin team in December 2018. It is an extension to Zerocoin protocol. With Lelantus, there is no need for a trusted setup anymore. This new approach hides the origin and the amount of coins in a transaction.

7.5.2 Zerocash Protocol

Zerocash algorithm [19] was introduced in 2014. It guarantees strong anonymity. It is a decentralized anonymous payment protocol. It uses zero-knowledge Succinct Non-interactive ARguments of Knowledge (zk-SNARKs).

Security is increased by using 128 bits in Zerocash. Zerocash is much faster than Zerocoin. Anonymous transactions of variable amounts are allowed. The system hides transaction amounts and the values of coins. Without user interaction, anyone can do payments to another's fixed address. Functionality and performance are significantly improved in Zerocash, so it is a strong alternative to traditional Bitcoin. For the personal data protection aspect, it is totally a better choice since it guarantees anonymization. The Zcash and zk-SNARK are given in the next topic which are based on Zerocash protocols.

7.5.3 Zcash and Zk-SNARK

We can anonymize not only the transactions but also the sender, the recipient and other details by using zero-knowledge proof (ZKP). In the previous section, Zerocoin and Zerocash are explained. Zcash [17] is a successfully implemented blockchain based on Zerocash.

Zcash uses a modified version of zero-knowledge proof called zk-SNARK. It totally provides privacy for users for It stands for 'Zero-Knowledge Succinct Non-Interactive Argument of Knowledge'. The main problems of ZKP are the size of computation time and amount of proofs for validation. The zk-SNARK reduces both of them. As a summary, zk-SNARK's process can be described as below (taken from [25]):

- "Disassemble the code into verifiable logical verification steps, then disassemble these steps into an arithmetic circuit consisting of addition, subtraction, multiplication, and division."
- "Conduct a series of transformations to convert the code to be verified into a polynomial equation, such as $t(x)h(x) = w(x)v(x)$."
- "In order to make the proof more concise, the verifier randomly selects several checkpoints, in advance to check whether the equations at these points are true."
- "By homomorphic encoding/encryption, the verifier does not know the actual input value when calculating the equation, but can still verify."

- "On the left and right hand sides of the equation, multiply a secret value k that is not equal to 0. When verifying that $(t(s)h(s)k)$ is equal to $(w(s)v(s)k)$, the specific $t(s)$, $h(s)$, $w(s)$ and $v(s)$ cannot be known, so the information can be protected."

zk-SNARK relies on a principle that the parameters of trust settings must remain secret. If one of these parameters is leaked, the system can be under attack. The possible solutions of this problem is creating a trusted execution environment. One of the extensions of the zk-SNARK is ZK-STARKs (Zero-Knowledge Scalable Transparent ARguments of Knowledge). According to the zk-STARK, it does not rely on any trust settings. However, it is not tested on systems and it is at an early stage. These improvements in the digital currency sector also needed in other sectors. To create a proper personal data protection system, we can use the same principles.

7.6 Conclusion

We have highlighted the benefits of blockchain on personal data protection. We summarized proposed models by using zero-knowledge proof which are Zerocoin and Zerocash. We expect this paper will be useful for people who concern personal data protection in practice, and for researchers in directing their future efforts. Blockchain technology is an important technology and has many advantages. It is simple, transparent and efficient among other personal data protection methods. Blockchain allows full control of the data. It is more secure than centralized systems. There are also disadvantages of using blockchain. It is still at a very early stage. It requires training since personal data protection is an interdisciplinary subject like law, computer, etc. Changing old software can be difficult and costly for everyone. Last but not least, there are insufficient developers to implement a blockchain and keep maintenance of it. Despite these disadvantages of blockchain present, it has a huge potential to change the world and complete the digital transformation on personal data protection.

References

1. Zyskind, G., Nathan, O., "Sandy" Pentland, A.: Decentralizing privacy: using blockchain to protect personal data. In: 2015 IEEE Security and Privacy Workshops, pp. 180–184
2. Nakamoto, S.: Bitcoin: a peer-to-peer electronic cash system. Consulted 1(2012), 28 (2008)
3. Liang, X., Shetty, S., Tosh, D.K., Kamhoua, C., Kwiat, K., Njilla, L.: ProvChain: a blockchain-based data provenance architecture in cloud environment with enhanced privacy and availability. In: 17th IEEE/ACM International Symposium on Cluster, Cloud and Grid Computing (2017)
4. Liang, X., Shetty, S., Tosh, D.K., Zhao, J., Li, D., Liu, J.: A reliable data provenance and privacy preservation architecture for business-driven cyber-physical systems using blockchain. Int. J. Inf. Secur. Priv. 12(4), 68–81 (2018)

5. Kuner, C., Cate, F., Lynskey, O., Millard, C., Ni Loideain, N., Svantesson, D.: Blockchain versus data protection. Int. Data Priv. Law, **8**(2), (2018)
6. Linn, L.A., Koo, M.B.: Blockchain for health data and its potential use in health IT and health care related research. In: ONC/NIST Use of Blockchain for Healthcare and Research Workshop. Gaithersburg, Maryland, United States: ONC/NIST (2016)
7. Vora, J., et al.: BHEEM: a blockchain-based framework for securing electronic health records. In: 2018 IEEE Globecom Workshops (GC Wkshps), pp. 1–6
8. Dagher, G.G., Mohler, J., Milojkovic, M., Marella, P.B.: Ancile: Privacy-Preserving Framework for Access Control and Interoperability of Electronic Health Records Using Blockchain Technology. Sustain. Cities Soc., vol. 39 (2018)
9. Amofa, S. et al.: A blockchain-based architecture framework for secure sharing of personal health data. In: 2018 IEEE 20th International Conference on e-Health Networking, Applications and Services (Healthcom), pp. 1–6
10. Ramachandran, A., Kantarcioglu, M.: Using blockchain and smart contracts for secure data provenance management. CoRR, abs/1709.10000 (2017)
11. Ramachandran, A., Kantarcioglu, M.: SmartProvenance: A Distributed, Blockchain Based DataProvenance System. CODASPY (2018)
12. Joshi, A.P., Han, M., Wang, Y.: A survey on security and privacy issues of blockchain technology. Math. Found. Comput. **1**(2), 121–147 (2018)
13. Vishwa, A., Hussain, F.K.: A blockchain based approach for multimedia privacy protection and provenance. In: 2018 IEEE Symposium Series on Computational Intelligence (SSCI), pp. 1941–1945
14. Jia, B., Zhou, T., Li, W., Liu, Z., Zhang, J.: A blockchain-based location privacy protection incentive mechanism in crowd sensing networks. Sensors **18**(11) (2018)
15. Feng, Q., He, D., Zeadally, S., Khan, M.K., Kumar, N.: A survey on privacy protection in blockchain system. J. Netw. Comput. Appl. **126** (2019)
16. https://zcoin.io/. Accessed on 30 July 2020
17. https://z.cash/ Accessed on 30 July 2020
18. Miers, I., Garman, C., Green, M., Rubin, A.D.: Zerocoin: anonymous distributed e-cash from bitcoin. In: 2013 IEEE Symposium on Security and Privacy, pp. 397–411. IEEE (2013)
19. Sasson, E.B., Chiesa, A., Garman, C., Green, M., Miers, I., Tromer, E., Virza, M.: Zerocash: decentralized anonymous payments from bitcoin. In: 2014 IEEE Symposium on Security and Privacy, pp. 459–474. IEEE (2014)
20. Kosba, A., Miller, A., Shi, E., Wen, Z., Papamanthou, C.: Hawk: The blockchain model of cryptography and privacy-preserving smart contracts. In: 2016 IEEE Symposium on Security and Privacy (SP), pp. 839–858. IEEE (2016)
21. Goldwasser, S., Micali, S., Rackoff, C.: The knowledge complexity of interactive proof-systems. SIAM J. Comput. **18**(1), 186–208 (1989)
22. Introduction to Zero Knowledge Proof: The protocol of next generation Blockchain. Ashish, 2018, https://medium.com/coinmonks/introduction-to-zero-knowledge-proof-the-protocol-of-next-generation-blockchain-305b2fc7f8e5. Accessed on 2 June 2019
23. Groth, J., Kohlweiss, M.: One-out-of-many proofs: or how to leak a secret and spend a coin. In: EUROCRYPT, vol. 9057 of LNCS. Springer (2015)
24. Jivanyan, A.: Lelantus: A New Design for Anonymous and Confidential Cryptocurrencies, December 2018. https://zcoin.io/papers/lelantusv2.pdf. Accessed on 30 July 2020
25. WTF is Zero-Knowledge Proof, Oscar W, 2019, https://medium.com/hackernoon/wtf-is-zero-knowledge-proof-be5b49735f27. Accessed on 2 June 2019

Seval Capraz is a Ph.D. student at Hacettepe University, Department of Computer Engineering, Ankara, Turkey. She has a B.Sc. degree at TOBB University of Economics and Technology, Ankara, Turkey in 2012 and M.Sc. degree at Middle East Technical University, Ankara, Turkey in Computer Science in 2016. She works at Aselsan Inc. as a senior software engineer. Her research areas are distributed and parallel computing, high performance computing with GPUs, big data, blockchain and cryptocurrencies.

Adnan Ozsoy is an assistant professor at the Department of Computer Engineering, Hacettepe University, Ankara, Turkey. He received his Ph.D. degree from School of Informatics and Computing of Indiana University, Bloomington, USA in 2014. He did his M.Sc. degree in Computer Science from University of Texas at Austin, USA in 2007, and his B.Sc. degree from Virginia Polytechnic Institute and State University, USA in 2005. His research interests include parallel programming, high performance computing with GPUs, string matching algorithms, big data problems, distributed systems, application parallelism, blockchain applications and cryptocurrencies.

Chapter 8
Design and Verification of Privacy Patterns for Business Process Models

Masoud Barati and Omer Rana

Abstract Business process models can involve numerous operational activities for collecting, processing, and exchanging personal user data. Such processes may involve activities that are executed over one or more cloud-based platforms. With an increase in the use of enterprise business processes, the right to data privacy has become a key challenge for developers of process models deployed over such cloud platforms. Design of privacy patterns that are compliant with modern data privacy regulations remains a challenge with increasing adoption of such approaches. One such legislation is the General Data Protection Regulation (GDPR) aiming to protect European citizens from privacy violations, especially for data processing activities hosted within Europe or involving data of European citizens. Blockchain and smart contract technologies have been identified as promising approaches for supporting compliance checking and trust in business processes that utilize a distributed set of activities. Blockchains enable verification of GDPR obligations in an automatic way without the need of a trusted third party. This chapter describes how smart contracts can be used to meet GDPR compliance verification using a number of privacy patterns for business process models. We also describe how a transition system-based automated tool can be used to support such verification. We conclude with a discussion of integrating automated compliance checking (especially for personal user data) and the potential impact this has on the overall execution performance of business processes.

8.1 Introduction

The stakeholders of cloud markets wishing to deploy new privacy-aware technologies require to understand the privacy characteristics and concerns of services delivered to their customers [1]. One of the main challenges that providers of such markets

M. Barati · O. Rana (✉)
School of Computer Science & Informatics, Cardiff University, Cardiff, UK
e-mail: ranaof@cardiff.ac.uk

M. Barati
e-mail: baratim@cardiff.ac.uk

© The Author(s), under exclusive license to Springer Nature Singapore Pte Ltd. 2021
S. Patnaik et al. (eds.), *Blockchain Technology and Innovations in Business Processes*,
Smart Innovation, Systems and Technologies 219,
https://doi.org/10.1007/978-981-33-6470-7_8

face is designing their processing operations and data handling practices to serve end users more effectively and securely protect sensitive data from the access of unknown third parties. Business process models are well-suited for expressing the sequence of processing activities carried out by providers on personal user data during the execution of services. Such models enable the designers of cloud systems to clearly specify what privacy policies must be taken into account for the activities involving in a business process that collects or manipulates user data. Developing privacy patterns within the visual notations of business processes can improve the transparency of interactions between process stakeholders (e.g., cloud vendors) and data protection officers. These patterns should presently comply with common data protection laws to be readily tracked by the officers across the world. One of such laws is General Data Protection Regulation (GDPR) that imposes a series of rules for the actors handling EU citizen personal data [2].

GDPR includes four key elements, namely a data subject, a data controller, a data joint controller, and a data processor. Data subject is recognized by an identifier that can be name, address, identification number, etc. Data controller is an actor who determines data usage purposes. If such purposes are jointly specified by more than one controller, the notion of data joint controller arises. Data processor on behalf of a data (joint) controller is responsible for processing personal data. The combination of GDPR and cloud-based business processes has enabled cloud customers to exert control over their data and inquire whether their data is processing in a lawful manner [3]. For instance, a formal representation for tracking GDPR obligations within the business process of cloud services was proposed in [4] under which any processing activity of provider is subject to receiving user consent. In [5], the integration of GDPR and business process models provided a privacy-aware solution and helped stakeholders to be informed about privacy risks at early stage of system development.

Aside from realizing GDPR requirements, cloud service developers have recently made use of Blockchain and smart contracts in their designed business process models to enhance security and trust of customers [6, 7]. A Blockchain is a public ledger that includes a shared and distributed database (blocks) together with a number of peer-to-peer nodes, also called *miners* [8]. Blocks contain valid transactions and are created by the miners through a mining process [9]. Smart contracts are executable codes running on Blockchain and refer to the translation of common contracts between parties into a computer program [10]. The integration of Blockchain and business processes has led to the emergence of promising cloud-based applications that provide the transparent interoperations of service vendors and improve the reliability of services [11]. In [12], business processes and smart contracts have been combined to ensure data confidentiality by encrypting-sensitive data during business process execution. A framework utilizing Blockchain was developed in [13] that enhanced trustworthiness of business processes. The framework defined a process for diagnosing problems, discovering solutions, and redesigning business processes.

Some recent approaches use both GDPR and Blockchain in the business processes of cloud and IoT-based systems in order to enable the audit trail of data processors and securely protect user data [14, 15]. In [16], the design patterns of cloud services were equipped with a number of smart contracts that supported GDPR requirements and enabled the verification of cloud providers in an automatic way. The authors in [17] integrated GDPR-based smart contracts into the business process of smart devices to verify some obligations (i.e., data protection, data minimisation, etc) over both devices and their administrators (data controllers). These approaches, however, did not provide a formal solution for checking the GDPR requirements that should be realized during the execution of business processes. In order to address this, we design and validate privacy-aware business process models meeting two GDPR principles, namely user consent and data accountability, at both design time and run time. We transform business processes into finite state machines to provide a design time verification under which the GDPR compliance of the processes are tested by a model checking tool. The runtime verification is undertaken through a smart contract that automatically reports violators who processed personal data without the awareness of users. The chapter by representing a cloud-based purchase order system evaluates the amount of memory, gas, and time used for verifying the system.

The rest of the chapter is structured as follows: Sect. 8.2 proposes the smart contracts supporting GDPR requirements and illustrates a cloud-based scenario. Section 8.3 verifies business process models in both design time and run time. Section 8.4 describes experimental results of our proposed verification, and finally, Sect. 8.5 concludes the chapter.

8.2 A Privacy-Aware System Using Blockchain

GDPR enforces cloud providers to clearly determine their purposes of data processing for end users [2]. The purposes of data processing or generally privacy policies are normally expressed by a number of processing activities such as access, store, and transfer. Such activities within a cloud-based system deal with some personal data, being necessary for handling or using a service published by the system. As a GDPR requirement, a provider only allows to collect or manipulate user data if a positive consent is received from the user. In order to build a GDPR compliance system for realizing such a requirement, Blockchain and smart contracts are presently integrating into the business process model of cloud services to provide users with data processing purposes in a legible format. Moreover, the contracts enable users to keep their positive/ negative consent as an immutable evidence in a Blockchain network that can be used for future verification. In addition to the user consent obligation in GDPR, data accountability of service providers is a principle in GDPR

under which any processing activity on user data must be transparent and comply with the purposes/ privacy policies already specified by provider. For realizing this principle in an automatic way, Blockchain and smart contracts recently appear in the design pattern of cloud-based systems to store some records required for tracking providers' operations on personal data. Each record shows what processing activity is executed by which provider on what data items. For verifying these records and also checking user consent, a new smart contract should be implemented to automatically flag any GDPR violation.

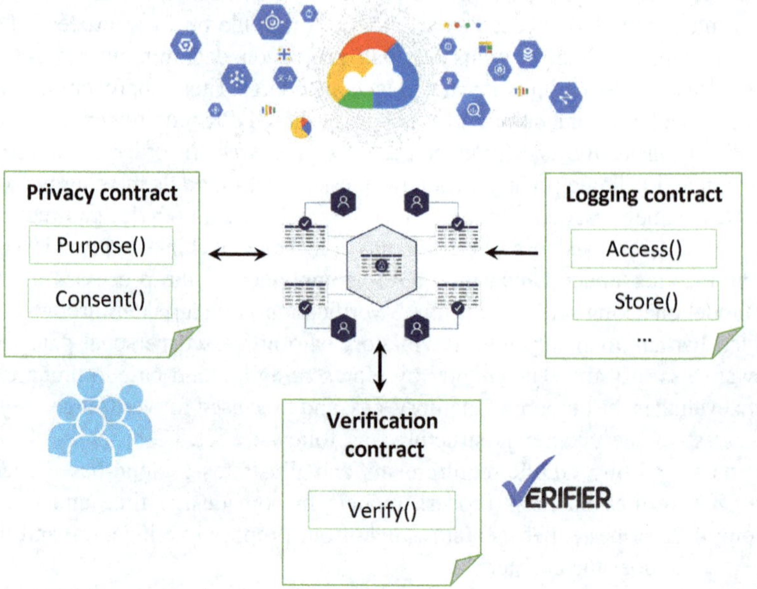

Fig. 8.1 GDPR-supported smart contracts

Figure 8.1 represents our proposed smart contracts for checking GDPR compliance of cloud providers.

Privacy contract enables providers to send their data usage purposes (i.e., what operations will be executed on what personal data items) in a Blockchain. Users through the contract can submit their votes, namely accepting or rejecting of data usage purposes, in the Blockchain. The contract activator can be both cloud users and providers, registered in a Blockchain network.

Logging contract through a trusted container hosted on a provider's side records: (i) provider address, (ii) processing activity (e.g., access, etc), (iii) processed personal data items (e.g., name, account number etc).

Verification contract retrieves the blocks already created by the privacy and logging contracts so as to detect the providers who breach GDPR obligations. The activator of the contract is a trusted party, called as *verifier*, connecting to the Blockchain network to report violators.

8.2.1 Scenario: A Cloud-Based Order System

Consider a cloud-based purchase order system in which customers demand goods through a portal, pay with their credit cards, and receive their orders through an online service delivery. The system also supplies a targeted marketing such as online

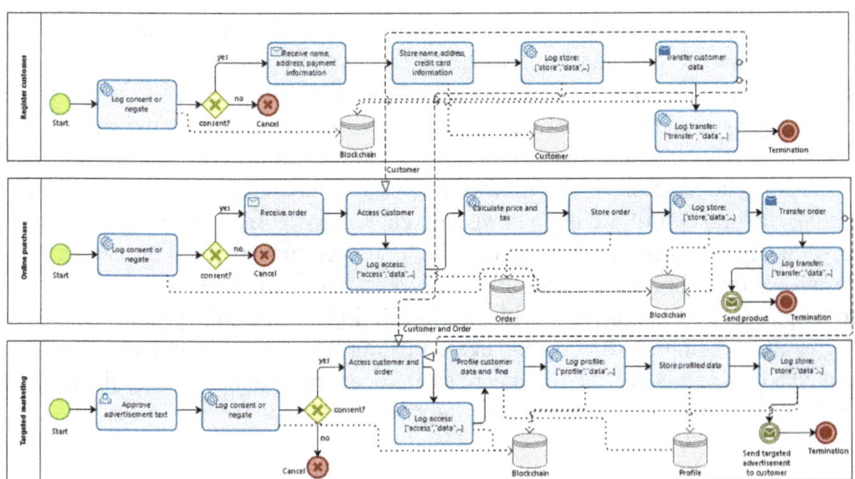

Fig. 8.2 Design pattern of a simple cloud-based purchase order system

advertisements to publish new offers for their subscribers. It is supposed that each typical service offered by the system is produced/ handled by a cloud provider: one for registering customer information, one for managing orders, and one for sending advertisements. The core business processes of such a system in BPMN representation, focusing on what personal data is collected and processed, is illustrated in Fig. 8.2. As seen, each process involves the functions of our proposed *privacy* and *logging* smart contracts that record some information in a Blockchain for enhancing data accountability in a transparent way. The *log consent* activity refers to a function in the privacy contract and *log access/ store /transfer/ profile* calls a function in the logging contract. The designed business process model for the order system is detailed as follows.

Register customer: A cloud customer signs up and subscribes with a portal, developed for online purchase. Through this business process, customers should provide a number of personal data, including email, postal address, and credit card information. The process contains data collection and storage operations. It also transfers personal data to other providers within the order system.

Online purchase: Subscribers can select their required products from a portal and make their payments using the credit card numbers already collected by the *register customer* process. The online purchase process calculates price and tax of selected products and directly submits such calculated data as an electronic receipt to customer. The process stores such receipts locally and transfers a copy of them to the *targeted marketing* provider for future automated processing operations.

Targeted marketing: uses the emails or mail addresses of customers to disseminate targeted advertisements based on the purchase history of customers. This process also makes an individual profile for each customer according their recent orders.

The processing activities of providers on user data within the business processes can generally be classified into: *access*, *store*, *transfer*, and *profile*. Precisely, the *access* or *data receive* activity is carried out by both online purchase and targeted marketing, as each of which should collect a part of personal data for offering services. The *store* activity appears within all business processes in order to keep user data (e.g., name, address, etc.), order information (e.g., purchased goods, buyer identification, etc.), and profiled data (e.g., age, purchased goods, etc.) in their local storage, respectively. The *transfer* activity is in both register customer and online purchase design patterns through which user data is sent to other processes. Note that *send* activity, shown by green circle with envelope in Fig. 8.2, delivers products or advertisements to customers directly and not to other providers; hence, it does not process personal data and is not classified here as a processing activity. Finally, the targeted marketing is only the process running the *profile* activity.

8.3 GDPR Verification

A business process model provides a visual representation for understanding what personal data manipulations are undertaken for delivering an online service to end users. In order to check the GDPR compliance of such data manipulations, two types of verification are presented: *design time* and *run time*. The former uses a model checking tool and enables system developers to verify a business process in accordance with GDPR requirements prior to its implementation. The latter makes use of a Blockchain-based technique for auditing trail of actor(s) during the execution of a service.

8.3.1 Design Time Verification

GDPR focuses on data processing purposes and encourages system designers to represent purposes in a legible and transparent form. The business processes involving in a process collection model can express the purposes of data processing [4]. In other words, a business process model can show the processing activities of cloud providers on personal data during the life cycle of service execution. For verifying such activities at design time, a formal representation is, however, required to ease checking the GDPR compliance of providers. Translating business processes into finite state machines is a solution for realizing such verification through existing model checking tools. Precisely, the business process of each service model using Blockchain contains a sequence of processing activities and also involves a set of activities running smart contracts (e.g., privacy and logging contracts) for tracking providers. The behavior of a service for executing such activities can be abstracted by an automaton. In some cases, a service provider may determine deadlines or time constraints for the execution of their activities. Take the retention period of data in the local storage of a provider as an example of such constraints. Thus, timed automata are appropriate means for supporting such constraints associated with the completion of activities.

A key element of a timed automaton is a set of clocks C, being non-negative real-valued variables. A conjunctive formula of terms in the form of $x \sim c$ or $y - x \sim c$, where $x, y \in C, c \in \mathbb{N}$, and $\sim \in \{\leq, <, =, >, \geq\}$, expresses a time constraint [19].

Definition 1 A timed automaton for a business process model is a tuple of $\langle \mathcal{A}, Q, q_0, F, C, \eta, H \rangle$, where $\mathcal{A}$ is a finite set of processing activities on personal data; Q is a finite set of states; $q_0 \in Q$ is the initial state denoting the start of a business process; $F \subseteq Q$ is the set of final states denoting the termination of data processing in a business process; C is a finite set of clocks; $\eta \subseteq Q \times \mathcal{A} \times G(C) \times 2^C \times Q$ is the transition relation, with $G(C)$ the set of constraints over C; and $H : Q \rightarrow G(C)$ is a function, which assigns an invariant to every state. The latter indicates the amount of time that may be spent in a state. The tuple $\langle q, \alpha, g, D, q' \rangle$ is a transition from q to q' on processing activity α with the clock constraint g. The clocks belonging to $D \subseteq C$ are reset to zero after taking a transition.

The processing activities in Definition 1 are those, directly processing or using personal data. Such activities can include the "collection (receiving), reading (access), recording, organization, structuring, storage, alteration, retrieval, dissemination, use, consultation, profiling, transfer, alignment, combination, restriction, and erasure of personal data" (Art. 4(2) of GDPR). Given the integration of Blockchain in our business process models, the processing activities also involve those running the smart contracts for logging the information required for tracking providers in a Blockchain. Generally, these information can be: user positive/negative consent, processed operation on user data, provider address, and processed personal data items.

Example 1 Consider the "online purchase" business process belonging to the design pattern of cloud-based purchase order system depicted in Fig. 8.2. The process involves the following processing activities set:

$$\mathcal{A} = \{positive\ consent,\ receive\ order,\ access\ customer\ data,\ log\ access,$$
$$store\ data,\ log\ store,\ transfer\ order,\ log\ transfer\}.$$

Assuming that, the provider of "online purchase" stores user data in its local storage for a minimum one hour and maximum two hours. Based on such a time constraint and the sequence of processing activities in the "online purchase" process, the timed automaton, depicted in Fig. 8.3, can be modeled for the process. As seen, the initial state is s_1, where the process start receiving user order. Each state shows the completion of a processing activity. Since there is a time constraint for *store data* according to the assumption, a clock x with the condition $60 \leq x \leq 120$ appears on its transition.[1] The *positive consent* activity refers to *log consent* in Fig. 8.2 with a condition that user accepts all the processing activities executed by provider on their personal data.

---

[1] The lower and upper bounds are in minutes.

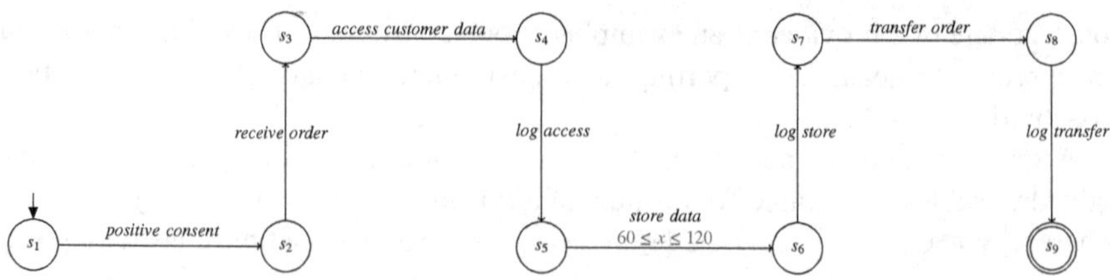

Fig. 8.3 A timed automaton for online purchase business process

In a GDPR-compliant business process model, our proposed logging smart contract should appear just after executing typical processing activities (e.g., access, store, etc.) to directly record the information which are necessary for verifying GDPR rules on providers. The privacy smart contract that logs user consent is located at the beginning of the business process model before any processing activity. Moreover, the retention period of data in providers' storage should be less than the total time, being necessary for processing personal data. To compare and test such a GDPR-compliant process model with those designed by system developers, we can use a model checking tool like Uppaal [18] that supports verification of timed automata. Using the behavior of a provider for processing personal data inside a business process, it is possible to model a specification as a desired timed automaton for the business process that complies with GDPR requirements. Then, the reachability of such a specification from its initial state to final state(s) can be compared with the timed automaton related to the actual version of the business process by defining a number of synchronization channels in Uppaal. If the interaction between the specification and actual automaton through the channel leads to a deadlock, it means that the latter is not compliant with GDPR.

A binary synchronization channel for a processing activity $\alpha \in \mathcal{A}$ is declared as "chan α" in Uppaal. The handshaking between the specification (sender) and actual automaton (receiver) via the channel of α is shown by "α!" on the transition of sending automaton and "α?" on the transition of receiving one. Each automaton can have its local clock. A clock is declared as "clock x" in the declaration part of the template related to the automaton. In Uppaal simulation, transitions can be taken in a manual mode (i.e., step-by-step) to validate the interactions and an automatic mode (i.e., a continuous way) to quickly identify possible deadlocks.

Example 2 Figure 8.4 shows the implementation of the automaton demonstrated in Fig. 8.3 in Uppaal. There are two automata, one for specification, the other for actual online purchase automaton (illustrated in Fig. 8.3). The interaction between the former and latter leads to a deadlock when the specification reaches state $t3$. Given the processing activities' set in Example 1, the receive_order belongs to $\mathcal{A}$ and a transition, conveying a smart contract (i.e., log_receive) for logging the activity in a Blockchain, must be taken from state $s3$. However, the actual automaton does not have such a transition. Hence, the automaton of online purchase violates data accountability principle in GDPR.

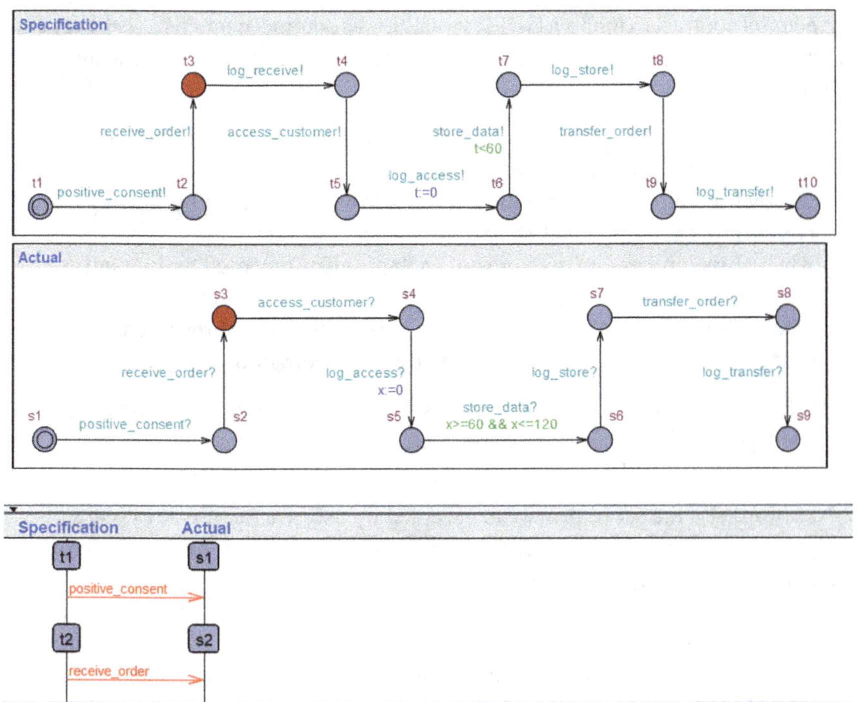

Fig. 8.4 Implementation of online purchase timed automaton in Uppaal

It is supposed that retention period for keeping personal data should be less than one hour (i.e., $t < 60$). Given this, another case for a deadlock in Fig. 8.4 is when the specification runs the `store_data` activity. To detect the possible deadlock, the following formula was tested:

```
A[] (Specification.t7 and Actual.s6) imply Actual.x < 60
```

The formula, however, was not satisfied, since the time constraint of actual automaton (i.e., $60 \leq x \leq 120$) violates the one in the specification, and hence, `store_data!` transition cannot be taken.

8.3.2   Runtime Verification

This phase of verification is undertaken after the execution of providers' activities on user data. For realizing such a verification, we implement a smart contract to automatically verify the information already recorded by the functions of *privacy* and *logging* contracts involving in the business process model. Verifier runs the *verification* contract to report a GDPR violator. The contract contains a *verify* function in which the Blockchains created by logging and privacy contracts are examined as follows:

1. whether the addresses of providers recorded by logging contract conform to those recorded via privacy contract or not;
2. whether the processing activities of each provider recorded by logging contract conform to those recorded via privacy contract or not;
3. whether the processing activities of each provider recorded by logging contract were already confirmed by user or not;
4. whether the personal data processed by each activity and recorded via logging contract conform to those claimed by privacy contract or not.

Based on the following notations, Algorithm 1 presents the verification of providers performed by a part of *Verify* function.

Notation: A_c: the set of providers accepted by user via the privacy contract;
A_e: the set of providers executed processing activities on personal data and recorded by the logging contract;
Op_a^c: the set of processing activities of provider $a \in A_c$ accepted by user via privacy contract;
Op_a^e: the set of processing activities executed by $a \in A_c$ on personal data and recorded via logging contract;
D_{op}^a: the set of personal data claimed by $a \in A_c$ for executing $op \in Op_a^c$ through the privacy contract;
$\mathcal{D}_{op}^a$: the set of personal data processed by $a \in A_c$ for executing processing activity op and recorded via logging contract.

Algorithm 1 The verification of providers

    Let $\mathcal{V}$ be a set containing violators' addresses
    Input: user address
    Output: $\mathcal{V}$

1: **function** VERIFY
2:     $\mathcal{V} \leftarrow \emptyset$;
3:     **if** $A_e \nsubseteq A_c$ **then**
4:         $\mathcal{V} \leftarrow \mathcal{V} \cup A_e \setminus A_c$;
5:     **for all** $a \in A_c$ **do**
6:         **if** $Op_a^e \nsubseteq Op_a^c$ **then**
7:             $\mathcal{V} \leftarrow \mathcal{V} \cup \{a\}$;
8:         **if** $\mathcal{D}_{op}^a \nsubseteq D_{op}^a$ **then**
9:             $\mathcal{V} \leftarrow \mathcal{V} \cup \{a\}$;
10:     **return** $\mathcal{V}$;

Table 8.1 Used memory and time for verification

Process	Register customer	Online purchase	Targeted marketing
Number of transitions	6	8	7
Number of time constraints	1	3	2
Used time for v_t (seconds)	0.002	0.005	0.003
Used time for v_d (seconds)	0.006	0.011	0.008
Used memory for v_t (KB)	26321	85272	57386
Used memory for v_d (KB)	27524	29008	28504

As seen from the algorithm, the verification is done in three levels: the providers processing personal data, the activities executed by each provider on personal data, and the personal data requested for each processing activity. A violation is detected if: (i) a provider accesses personal data without the confirmation of user; (ii) an accepted provider executes a processing activity already rejected by user; (iii) the personal data items processed by a provider is much more than those recorded via the privacy contract (Table 8.1).

8.4   Experimental Results

This section evaluates our presented verification at both design time and run time. The former uses a timed-supported model checking tool (Uppaal) to check the GDPR compliance of the cloud-based purchase order system. The latter makes use of a Blockchain test environment (Ropsten [20]) to assess the gas consumption of proposed smart contracts.

8.4.1   *Evaluation of Design Time Verification*

This experiment by implementing the timed automata of cloud-based order system in Uppaal investigates the time and memory taken for verifying possible deadlocks. It assumes that we have three automata each of which refers to a business process depicted in Fig. 8.2. Moreover, a specification determines one time constraint for a processing activity in *register customer* process, three time constraints for three activities in *online purchase* and two time constraints for two activities in *targeted marketing*. Two types of verification were undertaken in Uppaal:

- whether the time constraints of specification are satisfied by the actual automata of business processes or not?
- whether the interactions between specification and actual automata leads to a deadlock or not?

The latter ignores the time constraints imposed by specifications and only checks the states' reachability in both a specification and its associated actual automaton. For better illustration in our experiment, the former verification is denoted by v_t and the latter by v_d. Table 8.2 represents the used memory and time for the verification. The data are calculated after five times verification with different time constraints to calculate average results. Since the automaton of *online purchase* has more transitions compared to the other automata, more time and memory is used for its verification. Furthermore, when the number of time constraints imposed by a specification increases, the amount of memory and time required for the verification goes up gradually.

8.4.2 Evaluation of Runtime Verification

We implemented and deployed our proposed smart contracts in Ropsten, which is an Ethereum [22] public test network providing more details about miners and has a gas limit of 4712388 for running contracts. The amount of gas used for deploying *privacy*, *logging*, and *verification* contracts were, respectively, 1356961, 527497, and 1246963. The effects of changing the number of processing activities and providers on the transactions cost and mining time are assessed as follows.

(a) **Number of processing activities and transaction cost**: This experiment by changing the number of processing activities executed by a provider on user data investigates the costs that must be paid for executing the transactions of *privacy*, *logging*, and *verification* contracts. The number of processing activities, randomly selected among *access*,*store*, *transfer*, and *profile* in the cloud-based order system, is varied from one to nine. Each activity collects or uses a number of personal data, which is randomly changed between one and ten for each execution. Moreover, the rate of gas price is 5 *gwei*. The proposed contracts were deployed in Ropsten test network, and the average costs were calculated after the execution of contracts' functions with different parameters. Figure 8.5 shows the results, where x-axis and y-axis represent number of processing activities and transactions' costs in USD, respectively. As seen, when the number of activities rises, the cost increases steadily. Because the complexity of transactions in the privacy contract, determining data usage purposes and getting user consents, was much more than the other contracts, much more should be paid for running the contract compared to the *logging* and *verification*.

(b) **Verification cost under different gas prices**: The experiment by changing the rate of gas price evaluates the amount of gas used for verifying GDPR obligations. The number of service providers is varied from one to three in the cloud-based order system. It assumes that we have three groups of services[2]:

---

[2]The assumption is that each service is supplied by a cloud provider.

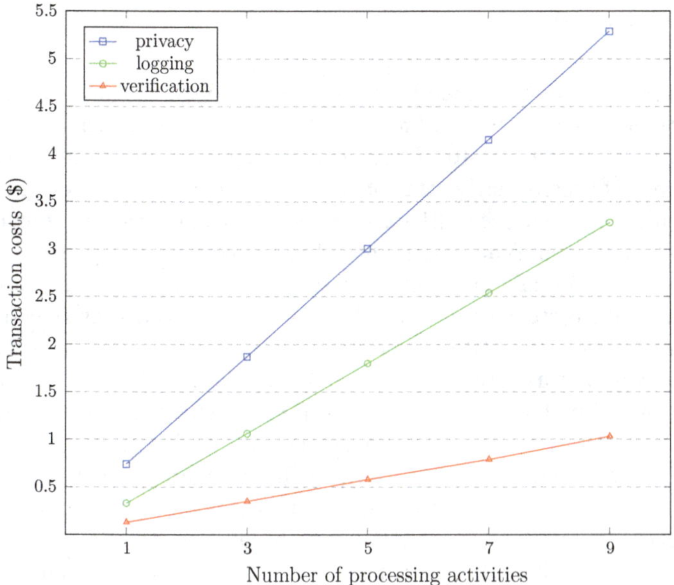

Fig. 8.5 Relationship between number of activities and cost of transactions

Table 8.2 Relationship between number of providers and cost

Service	g_1	g_2	g_3
Number of activities	3	6	9
Number of providers	1	2	3
Used gas	178333	349821	522502
Gas cost (gwei)	8916650	20989260	36575140
Verification cost ($)	3.98	9.03	14.30
Mining time (seconds)	455	205	33

g_1 : *{register customer} with* 50 *gwei gas price*

g_2 : *{register customer, online purchase} with* 60 *gwei gas price*

g_3 : *{register customer, online purchase, targeted marketing} with* 70 *gwei.*

Our proposed smart contracts were deployed in Ropsten test network, and Table 8.2 represents the verification costs and mining time. The gas cost in *gwei* is calculated as: *used gas × gas price*. As seen from the table, when the number of providers increases, the verification cost rises sharply. Furthermore, the average time for mining the verification contract decreases significantly when the rate of gas price increases. In fact, higher gas price rate motivates miners to accelerate mining procedures and quickly create blocks.

8.5 Conclusion

This chapter describes how cloud-based system developers can analyze GDPR compliance of their systems in both offline and online modes. The realization of the former was undertaken through transformation of business process models into timed transition systems to formally verify data accountability and user consent obligations. The online verification made use of Blockchain and smart contracts to automatically flag GDPR violations, identifying actors who collected and processed personal data without the consent of cloud users. A cloud-based purchase order system is use to show how smart contracts are integrated into the system's design patterns to record the key information required for GDPR-based verification, using a Blockchain. The cloud-based order system was verified by Uppaal in accordance with a GDPR-compliance specification and processing activities violating the specification were detected at design time. Compared with a similar approach presented in [21], the chapter evaluated the memory and time used for checking GDPR requirements in Uppaal. Moreover, we deployed the smart contracts involving business processes of the order system in Ropsten to assess their gas consumption. Finally, two different experiments were conducted to indicate how much cost, memory, and time are used for detecting GDPR violations in both online and offline modes. Future work focuses on translating other GDPR obligations, which can be prescribed for cloud ecosystems, into smart contracts. Furthermore, the scalability of our proposed verification remains another challenge for future investigation. However, a note of caution—there is significant operational and conceptual gap between law and technology. The "ways of working" of each field differ significantly. Data protection law concepts such as *personal data*, *processing*, *consent*, *purpose limitation*. and *legitimate interests*, to name a few, are overly broad, highly abstract, or involve an intricate substantive assessment. Computer science and software programming, on the other hand, rely on a concise language that fails to fully capture the nuances of legal clauses. Additional interdisciplinary research is required to broaden the set of legal rules, and support compliance verification that can be carried out through smart contracts.

Acknowledgements This work has been carried out in the "PACE: Privacy-aware Cloud Ecosystems" project, funded by EPSRC under project grant: EP/ R033439/1.

References

1. Russo, B., Valle, L., Bonzagni, G., Locatello, D., Pancaldi, M., Tosi, D.: Cloud computing and the new EU general data protection regulation. IEEE Cloud Comput. **5**(6), 58–68 (2018)
2. Virvou, M., Mougiakou, E.: Based on GDPR privacy in UML: Case of e-learning program. In: 8th International Conference on Information, Intelligence, Systems & Applications, Larnaca, Cyprus (2017)
3. Barati, M., Rana, O., Theodorakopoulos, G., Burnap, P.: Privacy-aware cloud ecosystems and GDPR compliance. In: 7th International Conference on Future Internet of Things and Cloud, pp. 117–124. Istanbul, Turkey (2019)

4. Basin, D., Debois, S., Hildebrandt, T.: On purpose and by necessity: Compliance under the GDPR. In: International Conference on Financial Cryptography and Data Security, pp. 20–37. Springer, Nieuwpoort, Curacao (2018)

5. Pullonen, P., Tom, J., Matulevicius, R., Toots, A.: Privacy-enhanced BPMN: enabling data privacy analysis in business processes models. Software & Syst. Model. **18**, 3235–3264 (2019)

6. Ulybyshev, D., Villarreal-Vasquez, M., Bhargava, B., Mani, G., Seaberg, S., Conoval, P., Pike, R., Kobes, J.: (WIP) Blockhub: Blockchain-based software development system for untrusted environments. In: 11th International Conference on Cloud Computing, pp. 582–585. San Francisco, CA (2018)

7. Barati, M., Petri, I., Rana, O.F.: Developing GDPR compliant user data policies for Internet of things. In: 12th IEEE/ACM International Conference on Utility and Cloud Computing, pp. 133–141. Auckland, New Zealand (2019)

8. Mohanta, B.K., Jena, D., Panda, S.S., Sobhanayak, S.: Blockchain technology: A survey on applications and security privacy Challenges. Internet of Things **8**, (2019)

9. Zheng, Z., Xie, S., Dai, H., Chen, X., Wang, H.: An overview of Blockchain technology: Architecture, consensus, and future trends. In: 6th International Congress on Big Data, pp. 557–564. Honolulu, USA (2017)

10. Mohanta, B. K., Panda, S. S., Jena, D.: An overview of smart Contract and use cases in Blockchain technology. In: 9th International Conference on Computing, Communication and Networking Technologies, Bangalore, India (2018)

11. Viriyasitavat, W., Xu, L.D., Bi, Z., Sapsomboon, A.: Blockchain-based business process management (BPM) framework for service composition in industry 4.0. J. Intell. Manuf. (2018). https://doi.org/10.1007/s10845-018-1422-y

12. Carminati, B., Rondanini, C., Ferrari, E.: Confidential business process execution on Blockchain. In: International Conference on Web Services, pp. 58–65. San Francisco, CA (2018)

13. Johng, H., Kim, D., Hill, T., Chung, L.: Using Blockchain to enhance the trustworthiness of business processes: A goal-oriented approach. International Conference on Services Computing, pp. 249–252. San Francisco, CA (2018)

14. Mohanta, B. K., Sahoo, A., Patel, S., Panda, S. S., Jena, D., Gountia, D.: DecAuth: Decentralized authentication scheme for IoT device using Ethereum Blockchain. In: IEEE Region 10 Conference (TENCON), Kochi, India, (2019)

15. Mohanta, B.K., Jena, D., Ramasubbareddy, S., Daneshmand, M., Gandomi, A.H.: Addressing security and privacy issues of IoT using Blockchain technology. IEEE Int. Things J. (2020). https://doi.org/10.1109/JIOT.2020.3008906

16. Barati, M., Rana, O.: Tracking GDPR compliance in cloud-based service delivery. IEEE Trans. Services Comput. (2020). https://doi.org/10.1109/TSC.2020.2999559

17. Barati, M., Rana, O., Petri, I., Theodorakopoulos, G.: GDPR compliance verification in Internet of things. IEEE Access **8**, 119697–119709 (2020)

18. Behrmann, G., David, A., Larsen, K.G.: 'A tutorial on Uppaal. In: Bernardo, M., Corradini, F. (eds.) Formal Methods for the Design of Real-Time Systems. Lecture Notes in Computer Science, Vol. 3826, pp. 200–236. Springer, Berlin (2004)

19. Alur, R., Dill, D.L.: A theory of timed automata. Theor. Comput. Sci. **126**, 183–235 (1994)

20. Ropsten testnet pow chain. https://github.com/ethereum/ropsten

21. Barati, M., Theodorakopoulos, G., Rana, O.: Automating GDPR compliance verification for cloud-hosted services. In: IEEE International Symposium on Networks, Computers and Communications, Montreal, Canada, (2020)

22. Wood, G.: Ethereum: A secure decentralised generalised transaction ledger. Ethereum Project Yellow Paper (2014)

Chapter 9
Blockchain Technology in Energy Field: Opportunities and Challenges

Fenhua Bai and Tao Shen

Abstract Recently, blockchains technique keeps attracting considerable attention from Finance and other fields due to its strong commercial viability. In accordance with substantial examples resulting from this context, it was identified as having the potential to create remarkable profit and innovation. So what could be the effect of the combination of blockchain and energy industry? First, we should analyse the current status and the trend of decentralization of the energy sector. Then, as a decentralized and data strong consistency technology, blockchain solutions could be adopted to address the existing pain spots in energy application. To our knowledge, the pivotal enabler of blockchains committed to security, trustworthiness and retroactivity for distributed renewable energy might be judged as decentralized consensus algorithm underpinning blockchain technologies. Consequently, a lot of consensus mechanisms may be proposed to optimize the energy transaction processing speed. Additionally, it becomes increasingly recognized that a successful energy trading market must be consumer-centric after going through the California power crisis. Therefore, the improvement of the participation of prosumers and the flexibility of the renewable energy sources transactive marketplace would be quite essential. To this end, leveraging blockchain, the reliability of information, distributed management, and carbon emission control can be accomplished. After elaborating the application scenarios of blockchain technology in the energy industry, the article ends with a discussion of its challenges to be solved in the future development, passing over the hype phase in the energy sector.

F. Bai (✉) · T. Shen
Faculty of Information Engineering and Automation, Kunming University of Science and Technology, Kunming, China
e-mail: 1023326472@qq.com

T. Shen
e-mail: shentao@kust.edu.cn

   141
S. Patnaik et al. (eds.), *Blockchain Technology and Innovations in Business Processes*,
Smart Innovation, Systems and Technologies 219,
https://doi.org/10.1007/978-981-33-6470-7_9

9.1 Introduction

Nowadays, distributed renewable energy sources (DES) (e.g., Wind and solar power) have attracted massive attention from numerous countries [1, 2]. According to the statistics, the annual power generation of DES is more than two trillion kwh, accounting for more than 85% of non-fossil energy in China. Owing to the characteristics of multiagent, distribution, and decentralization, the traditional centralized energy trading mode is inappropriate to be adopted for DES, which may result in additional costs and potential problems of poor efficiency [3, 4]. Furthermore, the volatility of DES caused by weather conditions is an unavoidable challenge for the management and operation of energy industry. Therefore, locally distributed control and management techniques are required to accommodate the decentralized marketplace. As a distributed ledger technology, blockchain paradigm is primarily proposed to deal with distributed transactions without central management [5–9]. Consequently, it can be not only used to guarantee the anonymity and reliability but also for decentralization and transparency, addressing the challenges faced by decentralized energy transaction, storage [10, 11] and management [7, 12–17] as well as other potential applications.

1. The pain points of energy industry

    (a) The power generation side

    From the perspective of different energy types, in comparison with fossil energy, the ratio of the renewable one will grow ever more rapidly. Considering energy production form, energy production would be inclined to gain more co-relation with distributed renewable energy on the basis of the original centralized conventional units, and distributed renewable energy may even become the main power source in the future. At the same time, the system scheduling operation will gradually move from centralized to distributed. However, along with the rapid expansion of new energy scale and the natural difficulty of management of DES, as well as the unique randomness, volatility, and low disturbance resistance, the operation risk will increase after its large-scale connection to the power grid.

    (b) The consumer terminal

    In the energy Internet value chain, consumers are both producers and consumers, namely prosumers. In the future, consumers can either generate electricity through their own distributed renewable energy or more importantly, provide user-side load resources to participate in demand response. Consumers can actively participate in the demand response of the community through the demand-side response plan, and can also join the project as a member of the virtual power plant. Concurrently, they can discharge electricity to the grid [1] through electric vehicles and energy storage facilities. Therefore, with their increasing importance in the energy value chain, how to break the existing one-way passive mode of energy production and consumption, how to effectively use a large number of intelligent electrical equipment deployed at the consumer end to participate in demand-side management [12–17] and protect the data privacy of consumers are urgent problems to be solved.

2. The inherent consistency of blockchain and energy Internet

In the energy Internet, relying on an open architecture, information can be accessed and obtained anytime and anywhere. Renewable energy, energy storage, and energy-using devices can be "plug and play". Blockchain technology is open (in the public chain) or partially open (in the consortium chain) [8] for the addition of nodes. Individuals or devices that join the blockchain system can participate in accounting and transactions.

Similar to the information sharing mechanism of social networks, in the energy Internet [18], energy exchange is carried out dynamically, nearby and in real time to achieve the optimal scheduling of global energy management with decentralized local optimization. In the blockchain, trust comes from the sharing of information between nodes. Through the form of distributed database, the entire system allows each participating node to obtain a copy of the complete database. In addition, the blockchain has built a complete set of protocol mechanisms, allowing each node of the entire network to participate in recording data while also to participate in verifying the correctness of the results recorded by other nodes. The data will be written into the block only when most of the nodes in the entire network (or even all nodes) confirm the correctness of the record. Therefore, the blockchain system can be regarded as a trust sharing system.

Compared with the top-down tree structure of the traditional power grid, the formation of the energy Internet is a peer-to-peer interconnection [16] among bottom-up energy autonomous units. For the blockchain system, the entire network does not have a centralized hardware facility or management organization, and the power as well as obligations are equal between any nodes, and the operation of the entire system will not be affected by a certain node being attacked or offline.

In the energy network, certainty, the energy production terminal and the consumption end not only should be interconnected but also be interconnected with the wide-area. Besides, different forms of energy can be converted and complemented each other to achieve optimal allocation of resources. Different from traditional network systems, the blockchain uses P2P technology. Hence with all nodes interconnected in pairs. Decentralization has been achieved in the absolute sense [5–7].

3. Motivations of applying blockchain in energy field

As aforementioned, energy Internet and blockchain technology are consistent inherently. With the increasing development of information technique, the production, transportation, storage, consumption and other links of energy industry value chain are rapidly upgrading from mechanical or analog equipment to digital devices. In the future, the equipment will present the development trend of universal intelligence and IoT. Smart contract blockchain-based can provide intelligent rules for the interaction between multiagent in energy industry. Furthermore, it can realize all kinds of complex logic functions without participation, called Turing completion. The real significance of the smart contract based on blockchain lies in the characteristics of unforgeability and collective consensus. The pre-written code can directly call the data on the blockchain without any intervention, perform all the logic functions that can be calculated and output the results.

Specifically, the homogeneity between blockchain and energy Internet is as follows:

Openness: In the energy Internet, relying on the open architecture, information can be accessed and obtained anytime and anywhere. Renewable energy, energy storage and energy consuming devices can be "plug and play". Blockchain technology is open (in the public chain) or partially open (in the alliance chain) for the joining of nodes. Individuals or devices added to the blockchain system can participate in bookkeeping and transactions.

Sharing: Similar to the information sharing mechanism of social network, in the energy Internet, the energy exchange is carried out in real-time nearby, and the optimal scheduling of global energy management is realized by decentralized local optimization. In blockchain, trust comes from information sharing between nodes. The whole system through the form of distributed database, so that each participating node can get a copy of the complete database. In addition, the blockchain constructs a whole set of protocol mechanisms, so that each node in the whole network participates in recording data, and also participates in verifying the correctness of other nodes' record results. Only when the most nodes (or even all nodes) of the whole network confirm the correctness of the record, the data can be written to the block. Therefore, blockchain system could be regarded as a trust sharing system.

9.2 Blockchain Solutions for Energy Internet

Because of the particularity and foundational nature of the energy industry, once erroneous data or even malicious data is adopted and used to make decisions, great losses will be irreparable. This is an important feature distinguishing the energy industry from other industries. Therefore, in the energy Internet system, the authenticity and reliability of data are crucial, even exceeding the economic indicators.

1. Authenticity and reliability of information

In the traditional energy system, the authenticity and reliability of data are decided by the third-party authoritative center institution. Once the credit, endorsement, or supervision of the authoritative central agency has been changed, it will be difficult to ensure the authenticity and reliability of data. However, in the concept of the energy Internet, the form of the energy industry will undergo a subversive change. A great number of the DES and new consumers integrating production and consumption will become important participants in the energy field. The interactions and transactions between these producers will no longer need to go through the central management organization. Therefore, a decentralized mechanism is required to ensure the reliability of information flow in the future distributed energy systems.

Blockchain, as an immutable, historical, and strongly endorsed distributed data architecture, participates in calculation and encryption through all verification nodes in the network, and validates the authenticity and validity of information based on the same consensus algorithm. After reaching the consensus [19–23] of the whole network, it is encrypted and written into the distributed sharing database. Trusted blockchain data based on the consensus of the whole network is tamper-proof and traceable. Blockchain enables trust and endorsement of the data quality in the decentralized energy Internet system. When system participants have questions about the authenticity of certain data, it is convenient to go back to the historical block records to check up whether the data information is correct, and to identify whether the data has been tampered or recorded incorrectly.

2. Information privacy management

In a conventional energy system, data and information are generally stored and used by the central management organization after collection. Since the central management agency can view any data of any user at any time, there is no privacy for energy consumers and producers. Moreover, the traditional central institutions generally do not open data to third-party service enterprise. As a result, a large amount of valuable business data cannot be utilized and the value of the data cannot be reflected. However, in case the central organization opens the data to the outside world, it will face the huge challenge of user privacy leakage.

3. Information trading and pricing

The information age is based on data, becoming the basis of energy Internet innovation. However, pricing [13, 24] and trading data have always been a major challenge for the energy Internet, mainly lying in two dilemmas. First, data confirmation right, that is, how to determine the right of data, not only the ownership of data but also the right of earnings and privacy. Then, the other one is data replication. As a special commodity, data is different from ordinary commodities in essence. Data exists in the digital space in the form of bits. There is little cost for copying or misappropriation. Therefore, how to protect the data from being copied and misappropriated during transmission is quite a difficult problem, indicating that the traditional commodity intermediary matchmaking method cannot meet the requirements of data sharing, exchanging, and trading.

4. Consensus algorithm

The power information data transmission represented by distributed energy transaction has the characteristics of real time, requiring high transmission speed for the selected blockchain technology. Because of its distributed architecture design, blockchain needs a consensus algorithm [19–23, 25–27] to coordinate the data consistency of all nodes in the whole network. The consensus algorithm is the core technology of blockchain, not only guaranteeing the security and stability of the blockchain but also as the critical factor affecting the operational efficiency of the whole blockchain system. At present, consensus mechanisms are commonly used in public blockchains such as PoW [19], PoS [20, 21] and DPoS [22]; applying in alliance chains include PBFT [23, 25]. What's more, the consensus algorithms improved including Reputation-based Byzantine Fault Tolerance algorithm (RBFT), Credit-delegated Byzantine Fault Tolerance algorithm (CDBFT), and Byzantine Raft algorithm (BRaft), etc., are proposed. Some others include Algorand [26] and Raft [27] are emerging. The complexity of the above consensus mechanism is usually polynomial or even exponential, making the consensus speed unable to meet the requirements of the energy trading field. Therefore, the further explorations and researches are needed for the high-performance blockchain consensus mechanism.

9.3 Application Scenarios of Blockchain Technology in the Energy Industry

1. Auxiliary services

At present, the compensation cost for auxiliary services is shared equally by all levels of grid connected power plants according to the generation capacity, and then the funds are paid to the units providing services. In general, ancillary services trading has established a new mechanism to encourage competition and promote the safe and stable operation of power grid, but there are still some problems. First, there is no auxiliary service dispatching mode oriented to maximize the comprehensive benefits of power system. Second, there is no market-oriented auxiliary service pricing standard. Finally, there is no market transmission mechanism of "who benefits, who pays". These problems not only limit the healthy development of ancillary service market but also make it difficult to cope with the power production challenges brought by large amount of distributed energy and renewable energy in the future.

Blockchain-based solutions can solve the above problems. Firstly, a blockchain network involving all power plants and some large loads is established owing to dispatching agencies, power grid operators and regulators joining the network as verification nodes. Accordingly, all power station operation parameters must be registered and verified in real time on the blockchain to ensure the comprehensive and true reflection of the whole network to system operation status.

2. Distributed operation and maintenance management

Although there is no technical difficulty in the operation and maintenance of distributed photovoltaic and wind power, the cost of DES is relatively high due to the geographical dispersion of distributed projects. Moreover, the availability is seriously affected because of the difficulty in finding and solving problems in time. The power plant owner, the third-party service enterprise of the power station, and the power plant equipment production company are taken as the nodes to construct the blockchain network. Then the production log of the power station is verified by consensus, and the asymmetric encryption is carried out by the public–private keys. Consequently, the authenticity, sequence continuity, and unforgeability of the generation data for the power station can be ensured while the real-time monitoring of the operation and maintenance of the power station is implemented. Once the power station quality problems occur, the operation and maintenance personnel can be automatically and quickly informed for maintenance.

3. DES generated electricity metering and subsidy management

As power supply and electricity consumption coexist in DES, there are various policy mechanisms for DES subsidy, which is not consistent with the price of electricity. Therefore, it becomes particularly important to accurately measure power generation and electricity consumption. By virtue of the blockchain, the meter's power generation measurement data can be recorded. Based on the blockchain storage and registration, the generation data collected from the electricity meters has the characteristics of unforgeability and consensus trust verified by each node. The DES department can design smart contracts with different generation subsidy schemes and automatically allocate subsidies to the DES owners according to the generation data recorded on the blockchain.

4. Carbon market application

Carbon emission right is a method to allocate and measure the carbon dioxide emissions of various industries. Those whose emission more than the quota will be punished. The extra carbon emissions need to be offset by additional purchase of carbon emission rights. The participants with balance of carbon emission rights can transfer the surplus part to the subject with excess emissions so as to obtain profits. The power system is the major carbon emission, and will become the active sector of carbon trading. Therefore, the carbon market needs an automatic and intelligent certification mechanism. Blockchain can provide an intelligent system platform for carbon emission certification and measurement. Exploiting blockchain technology, each enterprise can be given a proprietary digital identity of carbon emission rights, stamped and recorded on the blockchain. The greenhouse gas emissions of industries are updated on the blockchain in real time. According to the carbon emissions, the consumption of carbon emission rights is automatically confirmed by smart contract. Conducting carbon trading [28, 29], whenever the ownership of carbon emission rights is transferred, the transaction information has been recorded in the blockchain and cannot be tampered with. The blockchain system automatically fines the enterprises exceeding the emission standards.

9.4 Challenges

1. Low throughput

Nowadays, the computing power and response speed of blockchain technology cannot meet the demands of high-frequency transaction and deployment. In the application process, although the transaction speed has been improved by the consensus algorithm aforementioned, higher requirements have been put forward for the storage space, computing power, and response speed of the system with the connection of energy chains of many subsystems in the energy Internet. Different from the traditional distributed system, the processing performance of blockchain system cannot be extended by simply increasing the number of nodes, but largely depends on the processing capacity of a single one. Security and stability are the core elements of node performance. At present, there is still a big gap between the throughput of hundreds of times per second of blockchain and the high-frequency trading and deployment speed of power system.

2. Lack of Oracle mechanism

In order to interact with the real world, the smart contract system needs the encrypted signature information submitted by the external system. This system becomes the oracle. The Oracle is a trusted subject, signing information about the state of the external world, because the signature can clearly realize the Oracle, allowing certain smart contracts to respond to the uncertain external world. However, due to the immature Oracle mechanism, the key theft of blockchain will lead to those control of nodes on the blockchain and deliberately input those false data into the smart contract, causing the difficulty of distinguishing the authenticity of external data and instructions, and resulting in certain risks.

3. Supervise and standardize policies

Regulators under the blockchain technology will face massive data, theoretically, the whole network data will be open to regulators. As a result, regulators may suffer from a lack of precise access to search and use these data. At the same time, regulators also need to regulate the transparency of information, for example, what is private, what is public, and what is only available to regulators. In addition, the gradual popularization of blockchain technology in energy field will make a subversive impact on the traditional business model of public utilities, and its decentralized and autonomous characteristics will dilute concepts such as regulation. Therefore, it is necessary for the government to establish and improve the institutional norms and legal protection of the energy blockchain. Different regulatory and normative methods of the government will have different impacts on enterprises, and the regulatory measurements will become the policy challenges for enterprises.

4. The industry lacks standards

Currently, the concept of blockchain is not new, not only for the reason that the blockchain technology is still immature, but the concept of "energy and blockchain" is also more complex and in the discussion stage. In the future, how to establish industry standards will be a challenge for energy enterprises. For example, whether the application blockchain protocol for power generation or for power consumption is the same, what kind of protocol should be used in different scenarios, how to allocate profits, how to ensure the stability of power system, and so on. All these issues need to be explored by industry pioneers, and industry standards should be formulated according to actual problems. In turn, the formulation of industry standards will affect the researches' and applications' direction of enterprises, which may bring profound changes to the whole energy blockchain.

5. Business model is not clear

In terms of revenue, the traditional platform, with its centralized characteristics, collects a lot of information to provide for the market demanders, who promotes the completion of the transaction and extracts a certain fee from the transaction as the revenue source. Nevertheless, the energy blockchain is a decentralized platform, and it is impossible to charge a certain fee through platform transaction. Therefore, how to obtain income for enterprises has become an urgent problem to be addressed. From the perspective of cost, the energy blockchain needs the support of distributed grid, which will cause huge fixed costs including smart grid equipment, Internet of things appliances, power generation equipment, energy storage equipment, and so on. This preliminary equipment needs to spend a lot of money, and how to reduce costs should be explored by enterprises. On the whole, certain business model adopted by energy enterprises to maximize their own interests is the most basic problem to be faced.

9.5 Conclusions

At present, the development of blockchain has just stepped into the right track and starts to get the attention of the mainstream society. However, the unprecedented decentralization and de-trust function of blockchain technology has brought huge shock to the industry. The concept of blockchain industrial revolution has been looming. Through its mandatory trust and peer-to-peer interaction function, blockchain has radically changed the mode of power concentration and operation, as well as the distribution methods for the ways of production and its results. Moreover, through the subversion of both sides, the level of productivity has been greatly improved.

Blockchain technology has the capability to greatly change the world, being compared with the technological revolution of power and Internet. The idea of decentralization and disintermediation represented by blockchain is also the inevitable result of the development of information technology and Internet to a certain stage. The combination of blockchain and energy Internet maintains strong internal consistency. As a representative of the new generation of Internet technology, blockchain can be deeply integrated with the energy Internet in the application and function dimensions, so as to cope with the current application challenges of energy Internet, such as the dilemma of deployment, promotion, and demand.

As an emerging technology, blockchain still has many fundamental problems to be resolved. The current performance, security, scalability, and other indicators of blockchain technology contain many problems. In addition to technical indicators, there is a fundamental conflict between the centralization thought represented by blockchain and the existing energy system with concentration and strong centralization as the core. It is necessary to design a new regulatory system and business model based on the decentralized blockchain system. Therefore, deconstructing the energy system from the technical, commercial, and regulatory perspectives, and reconstructing the energy Internet based on the decentralized paradigm are an unavoidable fundamental challenges facing the applications of blockchain technology in the energy field, required to be considered and solved by all experts interested in the construction of energy Internet system.

References

1. Agung, A.A.G., Handayani, R.: Blockchain for smart grid. J. King Saud Univ. Comput. Inform. Sci. (2020). https://doi.org/10.1016/j.jksuci.2020.01.002
2. Andoni, M., et al.: Blockchain technology in the energy sector: a systematic review of challenges and opportunities. Renew. Sustain. Energy Rev. (2018). https://doi.org/10.1016/j.rser.2018.10.014
3. Bao, J., et al.: A survey of blockchain applications in the energy sector. IEEE Syst. J. **99**, 1–12 (2020)
4. Mollah, M.B. et al.: Blockchain for future smart grid: a comprehensive survey. IEEE Int. Things J. https://doi.org/10.1109/jiot.2020.2993601 (2020)
5. Zhumabekuly Aitzhan, N., Svetinovic, D.: Security and privacy in decentralized energy trading through multi-signatures, blockchain and anonymous messaging streams. IEEE Trans. Dependable Secure Comput. 840–852 (2016)
6. Morstyn, T., Teytelboym, A., Mcculloch, M.D.: Designing decentralized markets for distribution system flexibility. IEEE Trans. Power Syst. **34**(3), 2128–2139 (2018)
7. Bahrami, S., Amini, M.H., Shafie-khah, M., Catalao, J.P.S.: A decentralized electricity market scheme enabling demand response deployment. IEEE Trans. Power Syst. **33**(4), 4218–4227 (2018)
8. Zhetao, L., Jiawen, K., Rong, Y., Dongdong, Y., Qingyong, D.: Consortium blockchain for secure energy trading in industrial internet of things. IEEE Trans. Industr. Inf. (2017)
9. Khalid, R., Javaid, N., Almogren, A., Javed, M.U., Zuair, M.: A blockchain based load balancing in decentralized hybrid p2p energy trading market in smart grid. IEEE Access **99** (2020)
10. Farad (2017) Commoditising forward purchase contracts in ultra-capacitor intellectual property rights on Ethereum blockchain. Tech. rep., Farad. Accessed 15 June 2017
11. PROSUME, Prosume decentralising power White paper. https://prosume.io/white-paper/. Accessed 2 Nov 2017
12. Zhou, Z., Wang, B., Guo, Y., Zhang, Y.: Blockchain and computational intelligence inspired incentive-compatible demand response in internet of electric vehicles. IEEE Trans. Emer. Topics Comput. Intell. **3**(3), 205–216 (2019)
13. Haider, H.T., See, O.H., Elmenreich, W.: Residential demand response scheme based on adaptive consumption level pricing. Energy **113**, 301–308 (2016)

14. Cheng, L., Yu, T.: Game-theoretic approaches applied to transactions in the open and ever-growing electricity markets from the perspective of power demand response: an overview. IEEE Access **7**, 25727–25762 (2019)
15. Huang, H., Cai, Y., Xu, H., Yu, H.: A multiagent minority-game-based demand-response management of smart buildings toward peak load reduction. IEEE Trans. Comput. Aided Design Integr. Circ. Syst. **36**(4), 1–1 (2017)
16. Dang, C., Zhang, J., Kwong, C., Li, L.: Demand side load management for big industrial energy users under blockchain-based peer-to-peer electricity market. IEEE Trans. Smart Grid **10**(6), 6426–6435 (2019)
17. Huang, A.Q., Crow, M.L., Heydt, G.T., Zheng, J.P., Dale, S.J.: The future renewable electric energy delivery and management (FREEDM) system: the energy internet. Proc. IEEE, **99**(1), 133–148 (2011)
18. Baliga, A.: Understanding blockchain consensus models. Persistent **2017**(4), 2017 (2017)
19. L. Park, S. Lee, H. Chang. A sustainable home energy prosumer-chain methodology with energy tags over the blockchain. Sustainability **10**(3), Art. no. 658 (2018)
20. Cheng, S., Zeng, B., Huang, Y.: Research on application model of blockchain technology in distributed electricity market. In: Proc. IOP Conf. Series Earth Environ. Sci. **93**(012065) (2017)
21. European commission. climate action policies. 2030 framework for climate and energy policies, 2014. [Online]. Available: https://ec.europa.eu/clima/policies/strategies/2030/
22. Hua, S., Zhou, E., Pi, B., Sun, J., Nomura, Y., Kurihara, H.: Apply blockchain technology to electric vehicle battery refueling. In: Proc. 51st Hawaii Int. Conf. Syst. Sci. pp. 4494–4502 (2018)
23. Feng, Q.,He, D., Liu, Z., Wang, D., Choo, K.-K. R.: Multi-party signing protocol for the identity-based signature scheme in IEEE P1363 standard. IET Inf. Secur. **1**(99), 1–10 (2020)
24. Fadlullah, Z.M., Kato, N.: Game-theoretic coalition formulation strategy for reducing power loss in micro grids. Evolution of Smart Grids. Springer International Publishing (2015)
25. Kumari, S., Khan, M.K., Atiquzzaman, M.: User authentication schemes for wireless sensor networks: A review. Ad Hoc Netw. **27**, 159–194 (2015)
26. Kumari, S., Li, X., Wu, F., Das, A.K., Arshad, H., Khan, M.K.: A user friendly mutual authentication and key agreement scheme for wireless sensor networks using chaotic maps. Future Gener. Comput. Syst. **63**, 56–75 (2016)
27. Kumari, S., Li, X., Wu, F., Das, A.K., Choo, K.-K. R., Shen, J.: Design of a provably secure biometrics-based multi-cloud-server authentication scheme. Future Gener. Comput. Syst., **68**, 320–330 (2017)
28. "Global EV outlook 2019 - International Energy Agency," 2019. [Online]. Available: https://www.iea.org/publications/reports/globalevoutlook2019/
29. Khaqqi, K.N., Sikorski, J.J., Hadinoto, K., Kraft, M.: Incorporating seller/buyer reputation-based system in blockchain-enabled emission trading application. Appl. Energy **209**, 8–19 (2018)

Chapter 10
BlockChain Technology for Energy Transition

Mary Jean Bürer, Massimiliano Capezzali, Mauro Carpita, Matthieu De Lapparent, and Vincenzo Pallotta

Abstract In this chapter, we provide a framework for assessing the relevance and impact of BlockChain technologies for enabling the energy transition from both the production and consumption sides of the equation. We will assess the issues and current solutions to understand how BlockChain combined with Smart Contracting could be best adapted and combined to serve energy transition objectives. The assessment is taken from both the technology and business models perspectives. Our framework is designed as a guide for the adoption and deployment of BlockChain and Smart Contract solutions in the energy sector for common and new use cases by taking the perspective of multiple stakeholders such as consumers (either private and industrial), producers, operators and investors in microgrids, DSOs and TSOs. Moreover, we will include aspects related to regulation and policies. The proposed framework and assessment methodology are the synthesis of a study realized in Switzerland where several representatives of multiple stakeholders have been consulted via interviews and a workshop held in mid-2019.

10.1 Introduction

The energy industry is in a profound transition due to new regulations and technology advancements. This transition is motivated by the following factors:

- Diversity of energy sources
- New and different places of energy production
- Emerging opportunities for producing energy for self-consumption and for selling it to other consumers
- Energy storage technology
- Information and computing technology

M. J. Bürer · M. Capezzali · M. Carpita · M. De Lapparent · V. Pallotta (✉)
Applied Science University of Western Switzerland—HES-SO, Delémont, Switzerland
e-mail: vincenzo.pallotta@heig-vd.ch

School of Business and Engineering of Canton Vaud—HEIG-VD, Yverdon-les-Bains, Switzerland

S. Patnaik et al. (eds.), *Blockchain Technology and Innovations in Business Processes*,
Smart Innovation, Systems and Technologies 219,
https://doi.org/10.1007/978-981-33-6470-7_10

These important changes provide new opportunities for the players in the energy industry at different levels. New business will emerge based on the possibility for energy production, distribution and trading.

Moreover, policy-makers are concerned with the climate change induced by un-optimized energy production/distribution/consumption. Therefore, new constraints are already in place and expected in the near-future in order to deal with this global issue.

Information technology plays an important role in this evolution of the energy market because of mainly two factors:

1. Intensive data production and data accessibility
2. Distributed computing advances, which allow the design of complex systems.

The BlockChain (BC) and Smart Contracts (SC) technologies are extremely relevant for the second factor (and indirectly for the first one) because it enables the "Internet of Value" [45, 47, 52]. With these two (strictly related) technologies, it is (and will be) possible to build systems where not only information is exchanged but also assets. BC and SC are expected to disrupt the traditional financial systems but also every system where any kind of goods or commodities are exchanged and traded.

Currently, most of deployments of BC and SC have been done in the financial system [e.g., cryptocurrencies like BitCoin and investments like in the case of Initial Coin Offering (ICO) developments]. In the Energy Market; however, the adoption of those two potential disruptors seems still to be slower. Among several reasons, we can certainly affirm that two are very important:

1. *Regulations of the energy market*: the government bodies, at different levels, from local to global, have developed a complex regulatory system, which is also influenced by different lobbies.
2. *Complexity of the energy market*: With "complexity" we mostly refer to the variety of stakeholders with different and often conflicting interests, but also to the intrinsic complexity of the energy distribution networks.

Nevertheless, the enabling power of these new technologies is such that they are going to be adopted by the players in this industry, maybe in an anarchic way. BC and SC have the potential of enabling full decentralization in the management of power networks to the point that energy consumers and energy producers will be able to do business without intermediaries.

There are also big expectations from those technologies, based often on a misunderstanding of their features. In particular, the constraints imposed by the existing grid infrastructure cannot be ignored and must be considered in the design of the energy trading overlay. However, the existence of those constraints can be the opportunity for developing new *Business Models* that were unconceivable when these technologies were not available.

Information and communication technologies are usually innovation enablers. In the context of the energy market evolution towards Smart Grids, BlockChain and Smart Contracts technologies can contribute to this evolution provided that new business models for energy trading are developed and implemented correctly. These models should also take into due account important factors such as the *environment*, *society*, and most importantly the *market*. In this paper, we aim at a better understanding of the role of these technologies in enabling innovative business models for the Energy Market toward the so-called "Energy Transition". Moreover, we provide a framework and a methodology for eliciting relevant business opportunities by considering the constraints imposed by existing (and future) regulation as well as the needs of the participating stakeholders.

10.2 What Is the "Energy Transition"?

The energy transition is one of the central issues of the beginning of the twenty-first century. As such, the concept is not precisely defined as it encompasses a broad array of policies aiming at reducing the contribution of non-renewable or, at least, *GreenHouse Gas* (GHG)-emitting energy sources at regional, national and transnational levels, along with fostering energy efficiency measures, for example, in the residential and industrial sectors. The stated overall objective consists in decreasing global environmental impact of energy generation and consumption, with the main driver being the drastic decrease of carbon dioxide emissions in the atmosphere (at the EU level, see, for example, the Second Report on the State of the Energy Union [18]).

While seemingly homogeneous in the political and civic discourse, such policies differ vastly from one country to the other, in particular regarding the inclusion or exclusion of certain renewable technologies—such as biomass or geothermal [19, 28]. The potential role of nuclear energy within an energy transition is also widely debated since its non-renewable nature and the associated cumbersome waste treatment are put in perspective with respect to very limited impacts in terms of both GHG and fine particle emissions [22, 37]. Indeed, nuclear energy represents a major heterogeneity factor between countries and policies related to energy transition.

On the demand-side, energy transition underpins more "usual" energy efficiency measures such as buildings retrofitting or low-consumption electric appliances but also a certain number of more fundamental evolutions, for example, in the industrial or mobility sectors. Indeed, industry is rather swiftly moving toward production approaches inspired by circular economy, that include valorization of waste streams, or lifecycle analyses in the choice of materials and processes [11]. In the mobility sector, the penetration of electric vehicles—despite the debate regarding their environmental impact depending on the origin of the power and the battery typologies— has been faster than any prediction, while hydrogen is expected to strongly impact the heavy-duty transport sector already in the short term, not to mention LNG for both road and maritime freight [40]. Hence, the need for innovative policy frameworks and business models constitute a major endeavor, to accompany and foster such structural changes.

10.2.1 Business Models in the Energy Sector

During the next 5–15 years, a massive disruption across the entire *energy value chain* is expected, which will affect a broad range of stakeholders in the energy sector. Again, this is fueled by multilateral efforts such as those focused on *decarbonization* of the global economy to address climate change, and a shift towards an increasingly clean, intelligent, mobile, and distributed energy system. The linear value chain (centralized generation to end customers) is expected to become an "energy cloud" where the system will be more sustainable, highly digitalized, and dynamic.

So far, the existing business models in the energy industry are centralized. Burger and Luke [13], show that the most common structure is that of "brokerage", which is mostly based on a central entity that governs the transactional system as shown in Fig. 10.1.

In the specific case of renewable energies (e.g., solar power) shown in Fig. 10.2, the business models are currently substantially the same as in distribution of energy produced in power plant. The main reason is because the business transactions are managed by the energy grid operators.

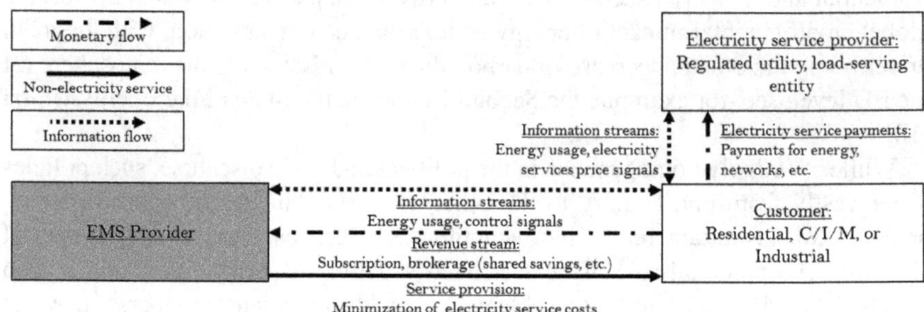

Fig. 10.1 Generic energy management service business model structure from Burger and Luke [13]

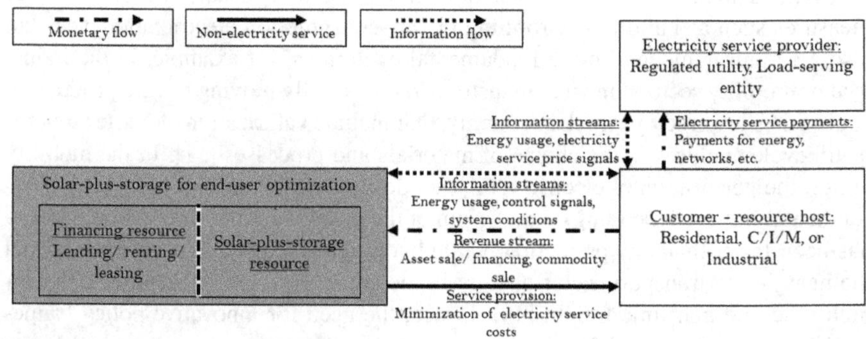

Fig. 10.2 Generic solar-plus-storage for end-user optimization business model structure from Burger and Luke [13]

Until now, innovative companies in the energy industry have focused mostly on grid management by integrating in a seamless way different sources of energy with the traditional grids by providing "smart" devices such as *smart meters*. The surplus of locally energy produced is typically "re-injected" into the grid and distributed to other consumers. The entity in charge of the distribution grid is also in charge of the billing and decides the prices. Prices are not traded among producers and consumers.

One of the simplest *value propositions* is that of the traditional energy supply business model. To remain profitable, national utilities rely on increasing kWh units sold [7, 26]. The ability of new entrants to compete in or join the market is affected by two key aspects of this business model (the national focus and relying on increasing unit sales) [25]. Furthermore, unsustainable practices are sometimes encouraged as a business model built on unit volume drives the whole energy value chain to increase throughput [4, 43, 48].

10.2.2 New Business Models for Energy Transition

It is well known that business models can emerge and change in industries in response to emerging technological change, technological opportunities, institutional change and pressures in the business environment. Disruptive and radical innovation [32], new technology paradigms [31], politico-institutional and socio-institutional dynamics [42] affecting a firm's external business environment, and more dynamics affect and create new markets that are disruptive to both producers and customers [34]. The role of national context for the emergence of new business models is also recognized [6, 12]. It is therefore clearly shown in the management literature that changes in the external environment can disrupt a firm's business model and open new opportunities for new business model design among start-ups and new entrants as well stimulate some level of business model reconfiguration among incumbents.

The diffusion of new technologies can also be spurred or supported by the catalyst of business model innovation by overcoming both internal and external barriers. The diffusion of sustainable innovations in particular has been linked to the implementation of new business models [9]. One can also imagine that business models that involve a network of players, creating a *value network effect*, can stimulate investments and diffusion of technologies that are particularly key to creating a successful complex innovation ecosystem. However, especially incumbents in an industry like the energy industry that have been operating for years with the same entrenched business model that is even protected by regulations in place today in most developed countries, may struggle to trust untraditional players that could create the right environment for such network value creation.

Complex *value creation* and *value monetization* among different *stakeholders* need to be better analyzed in such *value chain networks*. Increasingly, the value proposition of players in the energy sector involves previously ignored value propositions like environmental performance, affordability, reliability, perceived autonomy from the system, etc. And service providers find other sources of value creation in

B2B business opportunities, such as functionality, durability, reduced complexity, and attributes that might have been overlooked or less valued in the previous system. They may be seen as more important under a new paradigm such as responsibility of the provider. In addition, business models based on ownerless consumption can attract new customer segments, such as customers with lower financial capacity who may not be able to afford to purchase capital intensive technology [38].

Even though incumbents are hesitant to significantly move away from their formerly reliable business models, today the energy sector is in the midst of a major global transformation. While this transformation may occur more rapidly in the developing world where regulations do not protect large market players, and where many markets are still untapped by existing incumbents, this transformation is slowly happening even in the developed world despite the protection of the existing market in these countries.

This transformation is supported by multiple factors including:

1. many emerging technologies have already achieved critical mass, or it is on the immediate horizon;
2. innovative technologies are responding well to increased demand for new energy products and services;
3. viable business models are increasingly in competition with the traditional business models of the sector, which drive in turn incumbents to consider business model reconfiguration themselves.

According to Navigant, Customer-centric energy cloud platforms such as Building-to-Grid (B2G), Transportation-to-Grid (T2G), and Smart Cities will emerge little by little. In addition to this, energy carriers will become increasingly interconnected (Navigant, Energy Cloud 4.0, 2018). For example, hydrogen can be converted back into electricity and used directly as fuel for industries or transportation. Excess renewable power can be converted into heat, or hydrogen, and then transported, and stored. Other such applications will allow the transformation to be felt beyond the power grid. Finally, it will impact not only the way we live and work and move around but also the way we use materials, produce goods and transport them, as well as existing business models around services (Navigant, Energy Cloud 4.0, 2018).

New energy business models of the future will also deliver multiple benefits beyond the energy customer, toward the energy system itself, such as demand-side management reducing the need to reinforce networks [24], and fuel poverty alleviation [29] having public health benefits. Again, as mentioned above, carbon emission reduction is another value proposition or value capture element of many sustainable energy systems with wide benefits to society beyond customers. When new entrants, start-ups or incumbents begin to innovate business models in the energy sector including value capture for less tangible and easily tradable value propositions, complexity is certain in the business model development. Monitoring benefits accrued to different actors may be needed or capturing value from different actors could be needed (including from public sources) to compensate the enterprise or consumers for the "purchase" of such value which previously was ignored by the existing business model paradigms in the industry.

10.3   BlockChain and Smart Contracts in the Energy Sector

According to the Pw&C report on BlockChain technologies in the energy domain [50], the highest potential resides in the local renewable energy trading market. So far, the existing regulation is not adapted to the possibility of fully distributed energy systems (Microgrids or Smartgrids) where its governance is not centralized. Nowadays, apart from some episodic cases, the business models for microgrids do not yet leverage the BC and SC technology.

The challenges ahead for BC and SC technologies in the Energy Market are multiple especially in the context of energy transition. However, the possible disruptions might lead toward a truly liberalized and optimized market where the blocking factors in adopting renewable energies can be removed.

For BC and SC technologies, it is also fundamental to understand not only the technical aspects but also their impact in the existing systems, especially when the industry is extremely articulated and complex like the energy system.

The «use cases» proposed by Pw&C cover not only the obvious "billing" case, but also other important aspects of energy management such as:

- **Contracting** and enforcement of contractual terms;
- **Compliance** with existing regulations;
- **Transparency** in the transaction but also in data exchange and processing;
- **Resilience** of the network, which becomes very critical in a fully distributed one;
- **Liberalization** of the energy market with the creation of different interconnected "marketplaces" for energy trading.

Figure 10.3 compares the traditional structure of the energy market with that which might emerge from the adoption of BC and SC technologies. The removal of

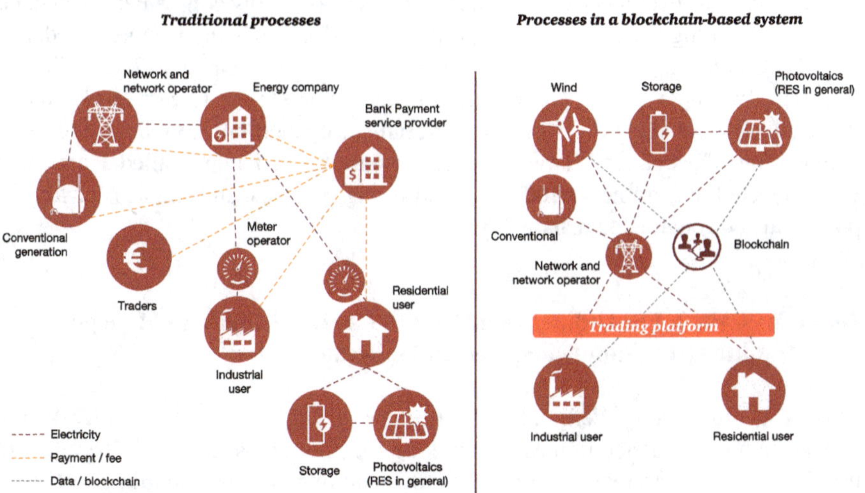

Fig. 10.3   Transformation of the energy market structure due to decentralization [50]

"clearing-houses" and the introduction of "trading platforms" are key elements of this disruption.

10.3.1 Types of Applications of BlockChain and Smart Contracts in the Energy Sector

In this section, we present specific use cases in the energy industry where BlockChain is being applied or has interesting potential to transform the industry.

10.3.1.1 Case 1. For Utilities that Want to Remain Competitive and Attract New Customers in a Liberalized Market

Blockchain applications are being developed by a number of players including grid management agents. Others are being developed to enable forecasting, or investments, or energy trade validation, or to trade green certificates or certificates of origin. Such features can improve the competitiveness of players in a more liberalized market scenario. It is worth noting that the Energy Web Foundation is working with electricity market participants from around the world to build a scalable, open-source BlockChain specifically tailored to energy market needs and designed to be energy efficient [49].

Singularity is also partnering with the Rocky Mountain Institute to establish an energy industry consortium with the goal of a more effective deployment of BlockChain to facilitate more effective operations in the energy sector. The new consortium aims to conduct R&D in BlockChain and energy in order to help utilities, application developers, customers, and renewable energy companies understand how the technology could support, disrupt, or transform existing business models.

BlockChain could indeed help utilities in some ways. For example, Tokyo Electric Power Co. wants to win back consumers, reversing an almost 15 percent decline in its customer base since the Japanese government opened up the industry to retail competition. The country's largest power provider formed a unit called Trende that will compete for customers with a solar and storage package and enable peer-to-peer power sales through BlockChain [35].

10.3.1.2 Case 2. For Utilities that Want to Use BlockChain to Manage the Grid (Supply and Demand in Real-Time)

This year, Burlington, Vermont, may become the first municipal utility to use BlockChain to get generation assets working together across its grid. The city will use the technology to manage supply and demand in real time, according to Killian Tobin, chief executive officer of Omega Grid, the BlockChain software provider that's helping Burlington set up its system. Think batteries charging when there's excess wind power, and businesses automatically drawing down power demand when electricity prices are high.

Efforts like this threaten to take command-and-control duties away from utility engineers and reduce the need for equipment upgrades that electric companies rely on for profits. "We're starting with a small utility and some microgrids, but we want to deploy on an entire grid," Tobin said [35].

10.3.1.3 Case 3. For Virtual Transmission—And a Move Away from the Usual Players

Germany's Tennet TSO GmbH is working with battery maker Sonnen GmbH and International Business Machines Corp. to form a virtual transmission line that uses BlockChain to store excess power from wind farms in thousands of home batteries in the northern part of the country and unleash power pent up in the south. Such efforts could eliminate the need for new power lines that utilities depend on for returns [35].

10.3.1.4 Case 4. For Peer-to-Peer Distributed Energy Trading Businesses

Peer-to-peer distributed energy trading (P2P DET) is the focus of a lot of BlockChain pilot projects. Lawrence Orsini, chief executive officer of LO3 Energy, is credited with facilitating the first peer-to-peer energy trade of solar power on a microgrid in Brooklyn, New York, in 2016. The same platform will also be used in Houston to help a group of businesses use their own resources to "micro-hedge" against swings in power prices instead of relying on a utility to do it for them [35].

BlockChain technology allows for the automatic matching of energy demand and supply via an online platform. Thus, homeowners who produce energy, most of the time through photovoltaic modules (PV modules), can sell their excess production to private individuals and households, usually located in the same geographical area. These transactions are validated and executed through smart contracts in the BlockChain. This kind of trading is called "peer-to-peer", since prosumers and consumers exchange energy directly between them without a middleman such as the utility company who manage the electrical grid.

Mengelkamp et al. [36] make a distinction between a virtual grid and a physical grid. The virtual grid is the information and communication network powered by BlockChain which connects all participants. The physical grid is the actual power distribution network [36].

As for the physical grid, a P2P DET system uses either the main utility grid or their own microgrid, which is "a self-sufficient local electricity supply system, either standalone or connected to a centralized grid of regional or national scale, comprising residential and other electric loads, and can be supported by high penetrations of local distributed renewables, other distributed energy, and demand-side resources" [23]. A microgrid is also called a "smart grid" because it is powered by digital communication technology which assists grid management.

Scholars [1, 41, 53] describe P2P DET architecture as typically composed of three levels:

1. The *microgrid*, which includes the prosumers, as well as the power generating and storage units.
2. The energy trading is carried out between microgrids within the same distribution network, also called a *cell*.
3. The energy trading can also occur between cells, namely at the *distribution system* level.

Moreover, energy can be exchanged not only on each of these three levels but also across levels (e.g., a prosumer in Migrogrid 1/Cell A selling his surplus energy to Cell B) [1, 41]. However, this architecture is only a theoretical representation. Apart from the first level, the distribution network and the distribution system level have not been implemented so far [1].

One point to clarify is whether the P2P DET system can be viewed as decentralized or distributed. It seems that there is no real consensus about this conceptual distinction in the literature. According to Gui and MacGill [23], a distributed *clean energy community* (CEC) is "a network of households and businesses that generate or own distributed generation individually, connected through a controlling entity either physically or virtually, and sharing the same rules in supplying and consuming electricity within the network". As for a decentralized CEC, it is "a community of households, businesses or a municipality that generates and consumes energy locally for self-sufficiency that may or may not connect to the main grid" [23]. Unlike distributed CEC, decentralized CEC is self-sufficient and independent from the centralized energy system [23].

Conversely, other scholars think that a distributed system allows for a greater freedom of its users than a decentralized system. In this perspective, the electricity and information flows go from the prosumers to the microgrid in decentralized systems while in a distributed system, they circulate directly between prosumers (Fig. 10.4; [3]).

This perspective is more in line with some classic approaches such as Baran's diagram in the field of communication systems. As Fig. 10.5 shows, a decentralized system is like a polycentric system while in a distributed system, there are no central points at all and each node is connected with all the others, whether directly or indirectly.

A distributed system corresponds more to a peer-to-peer trading system, where prosumers exchange energy directly between them. Consequently, we use the term "peer-to-peer distributed energy trading" (P2P DET) in the following sections.

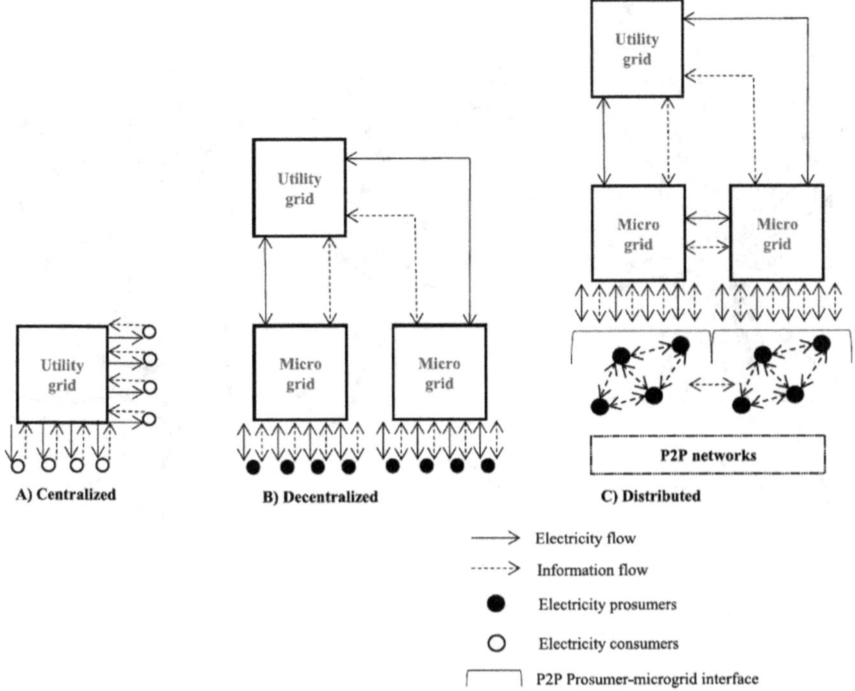

Fig. 10.4 Simplified overview of potential transition from **a** centralized, to **b** decentralized, to **c** distributed energy systems with microgrids and P2P networks [3]

10.3.2   New Business Models Enabled by BlockChain in the Energy Market

The Energy Market is likely to be transformed by BlockChain technology especially concerning financing models, contracting, and billing models. We examine in this section opportunities for each model.

10.3.2.1   Transforming the Financing Models

The financial sector is likely to be influenced significantly by BlockChain technologies in the future, and thus new opportunities will open up for new financing models relevant to the energy sector. Guaranteeing financial transactions and backing them against the BlockChain, and other use cases in the financial sector, is increasingly being embraced by innovative financial companies and could significantly impact the business models of most traditional financial institutions.

Many regions around the world have limited access to energy. BlockChains, combined with smart financing schemes, mobile applications, and digital sensors,

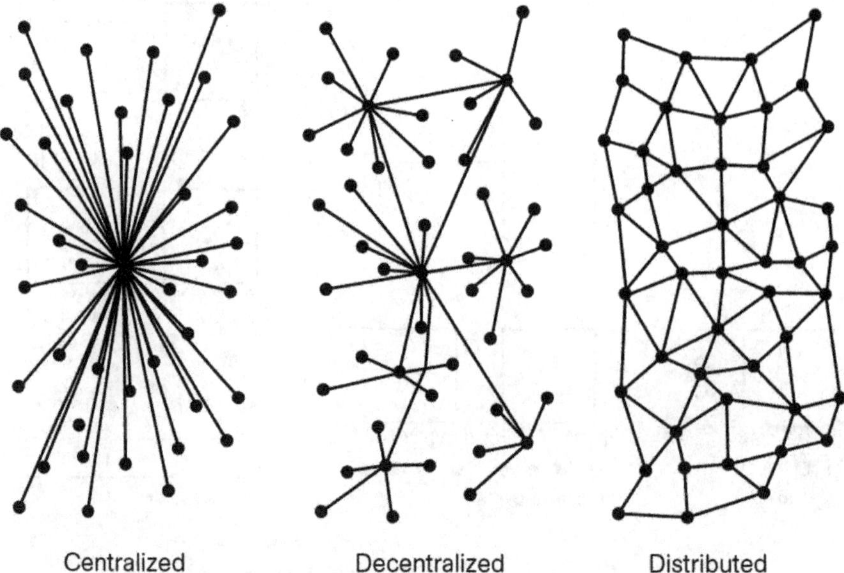

Fig. 10.5 Representations of centralized, decentralized, and distributed communications networks (Baran [5] in Gochenour [21])

can help distribute energy in small, discrete packets in these regions, allowing a local owner of a solar-generation system to sell power to neighbors. The solar-system owner installs a BlockChain-enabled solar panel on credit from the installer, using a mobile phone to pay for the hardware in installments and incurring minimal fees. Once the solar installation is paid for, the owner can sell small, discrete amounts of solar power to nearby consumers as they need energy. Power requests and payments can be made via mobile phone. The lighter fixed infrastructure involved with BlockChain and mobile micropayments allows these networks to thrive where other infrastructure—wires, traditional loan structures, and centralized energy authorities, for example—would be too cumbersome.

In one pioneering social initiative, the crowd-funding platform USIZO connected to BlockChain-enabled smart meters in underfunded South African schools so that donors can pay the school's electricity bills. BlockChain-based payments allow donors to ensure that 100% of each donation is used for its intended purpose. Similar methods can be used to provide electricity to new or underserved markets. M-PAYG, a Danish company, provides prepaid solar-energy systems to people living below the poverty line in developing markets and is leading a major project to electrify Uganda's largest refugee camp. For the power industry, the result is more individuals with power access and an increasing number of microgrids to support the main grid infrastructure. Owners of small solar-generation systems gain access to new income streams [27].

10.3.2.2 Transforming Carbon Management in the Energy Sector

The world of managing carbon emissions can also be transformed by BlockChain technologies. In a report by Infosys, [30], the BlockChain solution is laid out. It can be used as a tool to provide companies with readily available information for carbon emission calculations which is accurate, reliable, and standardized. The design and architecture of BlockChain technology promotes instant authentication, immutable data records and smart contracts. Those features allow for a fitting solution to integrate suppliers, manufacturers, logistics services providers, and stocking locations into a single network for rule-based interactions and value generation. Each party can report the carbon emissions across their individual value chains based on standardized metrics, creating a single platform for carbon measurement. Connecting BlockChain to various source systems such as ERP and SCADA will simplify data gathering across the supply chain, and improve the visibility at lower cost, time, and effort. BlockChain can provide a platform where every partner across the supply chain, i.e., manufacturers, suppliers, and distributors can work together in a transparent and accountable manner with the original equipment manufacturer or retailer to create a unified carbon ecosystem with more accurate measurement and credits [30].

10.3.2.3 Enabling Real-Time Transactions to Balance Supply and Demand

As solar and wind energy scale, power markets are increasingly challenged to balance supply and demand. Power supply was once provided by mostly "on call" or dispatchable sources of energy, such as coal and gas generation. In many markets, power supply varies with the wind and the sunshine. This has created demand for new "flexibility" services, to either adjust power demand to better match supply, or compensate backup sources of supply that can respond quickly in times of shortage.

As an example, UK-based Electron is using BlockChain to develop a platform for a flexibility marketplace, to allow real-time transactions to balance power supply and demand. This has been dubbed an "energy eBay," as it opens up participation in power markets. The trading platform would compensate consumers for adjusting their energy consumption, encouraging higher consumption in periods of high renewable power supply and lower consumption in periods of relatively low supply. It allows power generators and storage providers to transact in response to real-time price signals [27].

10.3.2.4 Managing Infrastructure in Real Time

BlockChain can enable more efficient monitoring and maintenance of power-industry infrastructure, based on secure, real-time data communicated by sensors. If an anomaly is detected, maintenance can be facilitated and paid for by smart contracts, leading to faster response times. Data is secure because it is only available to nodes in the BlockChain network. Again, BlockChain adds a layer of security and coordination to current digital pilots, enabling quick, accurate data gathering and communication between hardware suppliers, utility maintenance, and emergency response teams [27].

10.3.2.5   Connecting Electric Vehicle Charging Stations

In transport, BlockChain offers opportunities to coordinate electric vehicle (EV) charging. BlockChain facilitates energy payments at charging stations, allowing EV drivers to view maps of the charging network that highlight choices based on each user's preference and real-time pricing data. If BlockChain microgrids have been set up in the area, power prices at each station can be established by grid and residential power suppliers. Drivers can pay securely and instantly using a BlockChain wallet.

Using the Ethereum BlockChain to facilitate charging for electric vehicles is a project that was launched by German utility Innogy, assisted by a startup called Slock.it (Slock.it was since acquired), which specializes in providing BlockChain expertise to large corporations. They called the venture BlockCharge and promised seamless and affordable charging of electric vehicles. BlockCharge's business model was based on the one-time purchase of a Smart Plug and a micro-transaction fee for the charging process [46].

10.3.2.6   "De-comoditization" of the Electricity Markets

Another use case talked about on the Web is reinventing accounting systems to "de-commoditize" electricity markets. Leading energy companies are actively collaborating to create an open-source version of this application for global use within the Energy Web Foundation, a non-profit foundation focused on advancing BlockChain technology in the energy sector. In parallel, as we have seen early BlockChain applications are emerging in electric vehicle charging, microgrid management, and transactive energy. Automated demand response, wholesale market trading, and eventually fully distributed resource market automation are likely to follow. BlockChain is advancing at the speed of software development, not infrastructure deployment. There are many expectations in diverse industries ranging from banking to healthcare to sustainable agriculture. The hype about BlockChain has led to over $4 billion raised in 2017 via Initial Coin Offerings, largely unregulated vehicles for funding BlockChain-focused projects. However, for BlockChains to live up to expectations and achieve their full potential in the energy sector, several challenges need to be tackled [49].

10.3.2.7   The Peer-to-Peer Distributed Energy Trading Model

Pouttu et al. [41] have elaborated a theoretical and complex P2P trading model introducing a number of new actors and roles. Aside from the existing energy supplier, transmission system operator (TSO) and distribution system operator (DSO), new actors are the Microgrid Trader, who manages the commercial operations of the microgrid and trades energy with other microgrids, the Aggregator, which is a "legal entity that aggregates the load or generation of various demand and/or generation/production units to provide service to the wholesale market" [41], and the P2P-Trading Platform Operator who provides information, matches energy supply and demand, and possibly optimizes demand response at the cell level. This role can be combined with the Aggregator role [41].

A more in-depth analysis of business models enabled by BlockChain technology are presented in Bürer et al. [15].

10.3.3   Risks and Issues Related to BlockChain and Smart Contracts

BlockChain involves several risks that have to be carefully considered before promoted for system wide implementation, and in particular for the delivery of energy.

As for BlockChain technologies, in general, some standard risk considerations are: Strategic risk, Business continuity risk, Reputational risk, Information security risk, Regulatory risk, Operational and IT risks, Contractual risk, and Supplier risk. There are also several value transfer risk considerations, such as: Consensus protocol risk, Key management risk, Data confidentiality risk, and Liquidity risk. Finally, smart contract risk considerations include: Business and regulatory risks, Contract enforcement, Legal liability, and Information security risk [16].

BlockChain is also a platform for "smart contracts"—computer programs that automatically initiate certain actions when predefined conditions are met. The typical features of BlockChain technology such as redundancy, immutable storage, and encryption allow for decentralized validation. This decentralized validation allows for smart contracts. What is the difference between a "Smart contract" and a typical contract? A contract is an agreement having a lawful object entered into voluntarily by two or more parties, each of whom intends to create one or more legal obligations between them. Meanwhile, a smart contract is a computer protocol that facilitates, verifies, or enforces the negotiation or performance of a contract, or that obviate the need for a contractual clause.

Smart contracts can potentially encode complex business, financial, and legal arrangements in the BlockChain allowing for an infinite number of use cases and potentially beneficial outcomes for sustainability and a distributed income generation from new technologies such as renewables on household rooftops. However, there are risks to smart contracts as well. They can also result in the risk associated with the one-to-one mapping of these arrangements from the physical to the digital framework. In addition, cyber security risks increase as the smart contracts rely on outside oracles to trigger contract execution [16].

Technological superiority does not suffice, especially in the energy market. The risks associated with, or even just perceived to be associated with, BlockChain and its characteristics including anonymity which allowed it to develop for the cryptocurrency market, may be indeed the characteristics that block its development. Meanwhile, BlockChain is already transforming the financial industry and this alone will at least impact the financing models that apply to the energy sector. When BlockChain applications are increasingly applied in a secure and reliable way outside of the financial sector, it is possible that the energy sector will follow closely behind. But as discussed before the energy sector business models are difficult to change in most countries of the developed world today partly because of the protection created by regulatory frameworks created to protect consumers and industry which requires reliable electricity distribution and management.

Companies will face risks in their businesses when they must face completely new ways of doing business if they adopt BlockChain for more than superficial data management. They will have to decide on what business models they want, which partners they want to associate themselves with and rely on, and whether they are ready to manage a completely different business model such as one which is platform-based, or omni-channel based, and even if they want to apply BlockChain technologies to a product-based business model. BlockChain network governance will be key to manage the risks associated with mainstream use of BlockChain in any system, like the energy system.

10.3.3.1 BlockChain Technologies Impact in Power Distribution Grids

BlockChain technologies are considered as a promising route in order to facilitate the penetration of *renewable intermittent energy sources* (RIES) in particular, but not exclusively in urban zones, where the decentralized setting of RIES is indeed intrinsically structural. Indeed, the construction of BlockChain-based communication undoubtedly presents appealing characteristics. BlockChain can answer adequately to many challenges posed by the appearance of *prosumers*, along with the corresponding need to locally trade power production and consumption, also in the perspective of providing the necessary imbalance to the power grid. Massive penetration of RIES on any power distribution grid can potentially lead to transient overvoltages and crossing of current limits. Hence, any massive implementation of RIES must be accompanied by appropriate implementation of security components such as breakers, current limiters, grounding structures and, in the future, flexible SOP devices (possibly coupled with local storage capacities). The present security infrastructure of distribution grids cannot be considered as de facto sufficient: additional—potentially very relevant—costs must be accordingly evaluated and provisioned therefore in any BlockChain-based RIES development project.

In view of the above arguments, any future implementation of BlockChain infrastructures sustaining the deployment of RIES in any given territory should imperatively take these technical elements into account:

(a) Implementation of a wide-ranging power distribution grid simulation taking RIES penetration into account and determination of weaknesses on the basis of quantifiable indicators;

(b) Inclusion of real-time monitoring of the distribution grid, for example, by using technologies such as GridEye[17] by the grid owner/operator with possibilities to switch off generating capacities;

(c) Inclusion of grid management-related obligations and constraints in future smart contracts;

(d) Precise evaluation of additional CAPEX and OPEX costs related to distribution grid adaptations and potential reinforcements, including security;

(e) Distribution of above-mentioned costs among stakeholders, including utilities owning and/or managing the power distribution grid;

(f) Development of a sensible tariff system for grid usage that does not prevent the penetration of RIES, all the while covering all grid-related costs (maintenance, security, amortization).

In addition, implementation of BlockChain technologies in the power distribution system, even in dispersed and partial configurations (i.e., not covering the whole extension of a given distribution grid on a given territory) should probably be accompanied by a liberalization of the power metering market, which in most countries, is either completely under monopoly regime or only partially open to competitive bids.

In summary, whereas BlockChain technologies are able to provide a peer-to-peer energy trading framework favoring the implementation of RIES, while avoiding non-physical intermediaries such as banking agencies, promoters cannot avoid dealing with the physical intermediary constituted by the distribution grid itself. Hence, utilities must be embedded very much upfront in the any BlockChain-based RIES project and be considered as central stakeholders, to be involved in the overall value proposition.

10.4 The Framework

In this section, we present the outline of a framework and a methodology for guiding the adoption of BlockChain and Smart Contract technology in the energy market for enabling the Energy Transition goals.

Our framework is a blend of several methodologies in order to be adapted to the specific application to the Energy Market. We are taking it from the well-known Osterwalder and Pigneur's Business Model Canvas [39]. To illustrate the framework and the methodology, we discuss two use cases that were elicited during a multi-stakeholders Design Thinking workshop held in Switzerland in 2019, in which several local actors in the energy sector were invited.

10.4.1 The Business Model Canvas

The business model concept "was originally used to communicate complex business ideas to potential investors within a short time frame" [20]. Since then, the concept has developed into a tool not only for communicating the financial viability of a business project but also for analysis, comparison and performance assessment, management, and innovation [8]. A business model is particularly central to the firm competitive strategy. It helps defining the main characteristics of the goods or services proposed by the firm, the targeted market, the costs and revenues, the firm differentiation from the competition in terms of value proposition, and its integration in the value network [8]. In addition, business models act as catalysts for the diffusion of new and sustainable technologies [44]. They help overcoming internal barriers generated by the organizational culture and corporate governance (business rules, behavioral norms, performance indicators, and other control mechanisms) as well external barriers (high capital intensity, decision makers' reluctance to disruptive innovation) [9]. As such, "business models innovation dynamics are key drivers in accelerating the low carbon power system transition" [51]. Since the end of the 1990s, the business model approach has grown in popularity with the development of E-commerce and new ways of earning a profit through web-based products and services. The rise of sustainable technologies may have similar effects [9]. Nowadays, a widely adopted template in business and academia is the Business model canvas (BMC) proposed by Osterwalder et al. [39]. The BMC describes the value proposition, value creation and delivery, and value capture related to a product or service.

First, the value proposition dimension refers to the nature of the product or service offered by the firm (called "the value proposition" by the authors), the targeted customer segments and the type of customer relationship.

The **value proposition** *per se* "solves a customer problem or satisfies a customer need" and can be qualitative or quantitative [39]. Examples of value propositions are newness (satisfying new needs), better product or service performance, helping a customer get certain jobs done with a product or a service, design (e.g., fashion or consumer electronics industries), brand/status (e.g., Rolex), offering a similar value at a lower price, helping customers reduce costs, risk reduction (ex. service guarantee), accessibility, convenience/usability and customization («tailoring products and services to the specific needs of individual customers»).

The **customer segments** are of different types: mass market (one large group of customers with broadly similar needs and problems), niche market (specific, specialized customer segments), segmented (market segments with slightly different needs and problems), diversified (unrelated customer segments with very different needs and problems), multi-sided platforms or multi-sided markets (two or more interdependent customer segments).

Customer relationships connect the company with its customer segments. Their purposes are mainly customer acquisition and retention, sales-boosting but also raising customer awareness about the company's products and services, helping customers evaluate a company's value proposition, and providing the customers post-sale support. Categories of customer relationships are: personal assistance, dedicated long term personal assistance (e.g., private banking), self-service (no direct relationship with customers but the company provides all the necessary means for customers to help themselves), automated services (e.g., online services), user communities and co-creation of value with customers.

The second general dimension of the BMC deals with *value creation* and *value delivery*. It describes the production activities, the resources needed to produce the goods or services, the partners in the production process and the distribution channels.

As for **key activities**, they belong to three main categories: production (e.g., manufacturing firms), problem solving (activities such as knowledge management and continuous training, for example, consultancies, hospitals, and other service organizations) and platform/network (activities such as platform management, service provisioning, and platform promotion, for example, eBay, Visa, Microsoft).

Key resources are the most important assets in a business model. They are physical (e.g., means of production, infrastructures, distribution networks), intellectual (e.g., brands, intellectual property, partnerships, customer databases), human (skills, experience, knowledge) and financial (capital, funds).

Organizations conclude strategic **partnerships** with the purposes of optimizing the allocation of resources and activities, achieving economies of scale, reducing risk and uncertainty, and acquiring resources and perform activities. There are four main types of partnerships: alliances between non-competitors, between competitors ("coopetition"), joint ventures, and buyer-supplier relationships.

Channels are used by the company for communication, distribution, and sales to its customers. They are direct (e.g., online sales) or indirect (e.g., stores), company-owned or partner-owned (e.g., wholesaler). They serve two main functions: allowing customers to purchase the products and services, delivering a value proposition to customers through the adequate logistics or online infrastructure.

The *value capture* is the last general dimension of the BMC. It concerns the way the firm can turn part of the delivered value into economic profits, by considering its business costs and revenues.

Revenue streams are generated in several ways. The most common, asset sale, consists of selling ownership rights to a physical product. Usage fees are paid when using a particular service (e.g., telecom operator, hotel). The more a service is used, the more the customer pays. Subscription fees give a continuous access to a service (e.g., gym). Lending, renting, and leasing grant temporarily a customer the right to use a particular asset (e.g., car rental). Paying licensing fees gives permission to use protected intellectual property (e.g., media, technology). Brokerage fees are paid to intermediation services acting on behalf of other parties (e.g., credit card providers, real estate agents). Finally, advertising fees are charged for product and service advertising (e.g., media, event organizers). These different revenue streams are subjects to different pricing mechanisms: fixed "menu" pricing (predefined prices based on static variables), list price (fixed prices for individual products, services, or other value propositions).

Pricing depends on product feature, customer segment, volume (quantity purchased), market conditions ("dynamic pricing"), negotiation, inventory and time of purchase ("yield management"; for example, hotel rooms, airline seats), supply and demand ("real-time-market"), or competitive bidding ("auctions").

The *revenue streams structure* represents what is usually referred as the "business model" in the classic management literature.

The last building block of the business model canvas is the **cost structure**. Cost-driven business models aim at minimizing costs through, namely maximum automation and extensive outsourcing, and at proposing low prices (e.g., low-cost airline companies). Conversely, value-driven business models focus more on value and less on costs (e.g., luxury goods and services). Costs can depend on the volume of goods or services produced (variable costs) or not (fixed costs). Cost advantages can be achieved through economies of scale (reduction of the average cost per unit as the volume of output increases) and economies of scope (reduction of the average cost per unit due to product diversification) [39].

With the standard Business Model Canvas, we are able to represent and possibly design multi-party businesses with intricate relationships between different value propositions delivered to different customer segments through various channels and with a complex revenue streams architecture where it is not always the case that who benefits from the delivered value proposition is the one who pays for it. Also, there is some intangible value that is created through the business that cannot be matched by revenue streams, being this value delivered over a long term to someone who is not necessary part of a targeted customer segment.

This might be the case in the energy market and specifically in this delicate phase of energy transition where we need to consider social and environmental value created by the adoption of renewable energy sources. If we stick with ordinary business models where only tangible, often short-term value transfer is considered, we lose a large part of the picture. For instance, those partners who would help in building the grid infrastructure for renewable energy might be moved by environmental concerns rather than the pursuit of profits.

Being aware of this aspect, we could consider extending the Business Model Canvas with other components to incorporate those aspects by adopting a more sophisticated business model framework such as that of Joyce and Paquin [33] called the Three Layers Business Model Canvas (TLBMC) where in addition to the economic layer, the environmental and social layer have been added. When taken together, the three layers of the TLBMC provide a comprehensive insight into the economic, environmental, and social values created by the business.

For the purpose of the framework and the methodology, we decided to keep it simple and stick with the most intuitive framework of the Business Model Canvas. We went a step further and decided to simplify and focus it on essential elements that are relevant to the energy market. As shown in Fig. 10.6, we used a version of the Business Model Canvas where we took away the Customer Relationships, the Key Resources, and Key Activities from the original canvas. The main reason was to let the participants to the workshop to focus on the core aspects of the business model.

Fig. 10.6 Simplified business model canvas

10.4.2   The Use Case Selection

Before presenting the canvas to the participants, we created two teams and asked them to select a relevant use case based on two main challenges in energy transition described in Table 10.1.

Table 10.1 Use case description

Global context	Liberalization of the energy sector and new market opportunities within the framework of strong decentralization of energy. Currently, new sources of renewable energy are heavily subsidized around the world, but these subsidies are slowly disappearing. It will be necessary to have systems capable of accurately accounting for renewable energy production at multiple and dispersed production sites. Existing certification systems need to be improved if the system evolves toward a completely free market space, and also, ultimately, if the market evolves toward a scenario of total market liberalization for all segments (commercial and residential)
Local context	In the current market context, there is now an option supported by a new regulatory framework for microgrids for Swiss investors—the possibility of grouping together consumers or prosumers of renewable energy and gaining autonomy. This could apply to *microgrids* or eco-villages development projects, but also to different types of projects such as social or normal residential housing (new or existing projects), industrial parks, business parks, etc. Until now, these developments have been less attractive due to regulatory limitations on energy. Now, new business models are possible, and new technologies like smart grid or microgrid control technologies can allow optimization and integration with external power grids, thus providing a more interesting investment story for these developments

During the brainstorming, it should be noted that the moderators asked participants to think about the applications where BlockChain and smart contracting was needed for the given contexts. Therefore, a focus was put on where BlockChain could add value to the objectives laid out by some of the participants.

10.4.2.1 The Use Case for the Global Context: "Negawatt"

The workshop was seen by participants as a rare opportunity to think about uses for BlockChain that have not already been hashed-out and where value to society could lie. It was clear that from the brainstorming session of the Global Context group that the trading of guarantees of origin for green energy received more votes, however, the group discussed after the voting and decided that such green certificate trading had been explored more, but there were opportunities for Negawatt trading facilitated by BlockChain which had not been explored enough. They therefore chose this use case in the end for the following steps, because after the discussion, they viewed it as having higher value added by the group if solutions could be found to make the trading of Negawatts via BlockChain a reality.

Furthermore, the group interpreted the goal of this step as primarily looking for where BlockChain could add the highest value to the given goals, and it seemed that BlockChain and Smart Contracting could offer higher value to Negawatt trading than it could to the existing system of guarantees of origin or green certificate trading systems that already exist around the world under different policy initiatives and private-sector led voluntary programs.

The proposals are summarized in Fig. 10.7 and also grouped according two dimensions, namely Technology Added Value and Society Needs in Fig. 10.8.

National or global use cases brainstorming

	Negawatts certificates and/or contracts	Exchange platform for trading	Certificates or Gaurantees of origin for RE	CO2 credits and tracking CO2	Financing
G O A L S	Increased incentive for energy savings	EU cross-border trading possible	Local RE energy premium price because time, place, producer info in BC	Tracking CO2 on local level - Climate action and climate inaction made clear	Support for cooperatives (e.g. financing wind energy)
	Smart contracts faciliate exchange for smaller sources	Integrate cooperatives in the EU market	Certified identification of actors, sources, and quality	Incentivize various forms of CO2 reductions locally (CO2 credits)	Renter model for financing on a national level
		Investor security of service e.g. for foriegn sources	Long-term green energy contracts for producers (w/energy)		Crowd-funding facilitation
		Reduce barriers 4 entry into the market if liberalization	Production and storage certificates at EU level ; Virtual storage too		

Fig. 10.7 Results of the global use case brainstorming with the selected case

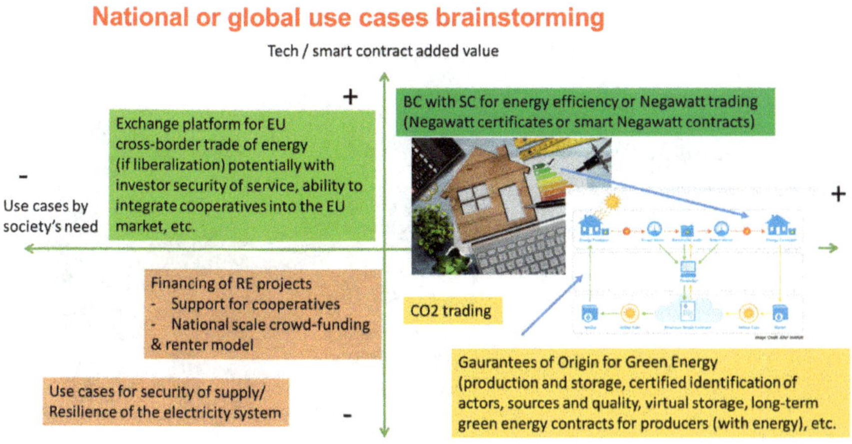

National or global use cases brainstorming

Tech / smart contract added value

Fig. 10.8 Global use cases brainstorming grouped by society's need or by technology value added

10.4.2.2 The Use Case for Local Context: "Dynamic Pricing for Microgrids"

The selection of the use case for the local context was easier as there was already a strong consensus on the fact that BlockChain would be the enabling technology for peer-to-peer energy trading in microgrids. However, some new elements were added because of the interaction between representatives of different actors in the energy market. For example, DSO representatives would argue that revenue streams could also come from selling microgrid management services or from selling collected data to companies for different purposes. Figure 10.9 summarizes the proposals and

Local use cases brainstorming

	Microgrid & RCP (contracting, measuring, and payments)	Financing/ Dynamic pricing	CO2/green certificates for local markets	Rent capacity of batteries/ demand response	One to peer concept	V2G and other management of charging of EVs
G	Quality (and autonomous) microgrids	Way to finance expensive equipment	Gaurantees of origin for decentralized production, RE	If resident is more energy effcient, they are compensated	Exchange from one PV producer to another outside	Facilitate payments in general at place of charging the vehicles
O	Managing members of RCP, authorizing contracts, etc.	Facilitate collective investment	Tracking of electric bicycle use, etc. possible	Can manage industrial site with residential site		Makes easier the exchanges between Evs and the grid
A	Services can be sold to the grid	Variable price of energy possible		Industrial ecology		
L	Makes financial exchange simple	Micro-transactions		Urban industrial design		
S	Flexibility market option	Dynamic pricing				

Fig. 10.9 Results of the local use case brainstorming with the selected (merged) case

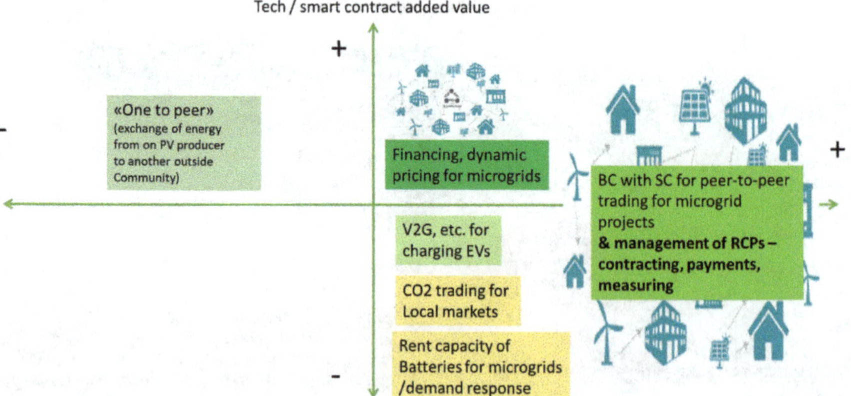

Fig. 10.10 Local use cases brainstorming grouped by society's need or by technology value added

Fig. 10.10 provide the grouping of the proposals according to Technology Value Added and Society Needs dimensions.

10.4.3 Impact Analysis of the Two Selected Use Cases

After the workshop, our team grouped the various use cases into categories as per two key criteria:

1. the *usefulness of the technology* toward the issue at hand or the value added potential from the technology proposed, compared to other technological options, and

2. the *importance of the case to society* (although this depends on the perspective one has about what transition scenario is most likely to offer benefits to society).

From this analysis, we see part of the reason why the groups chose the use cases that followed into the ideation and prototype steps which will be explained in the following chapters.

If we consider different perspectives about what is important to society, we can understand the value of each chosen use case compared to the other (Figs. 10.11 and 10.12). In short, blockchain with smart contracts is more useful in the case of energy efficiency trading (Negawatts trading) than in the case of microgrid projects with peer-to-peer trading however the peer-to-peer trading opportunities that blockchain might bring about eventually have greater importance to the energy transition if one believes in the decentralized energy transition scenario. However, if one assumes that energy efficiency gains are more important for an energy transition to be truly successful, then the Negawatt trading use case is the winner between the two options for investment, because blockchain is relatively more powerful in changing the dynamics of the

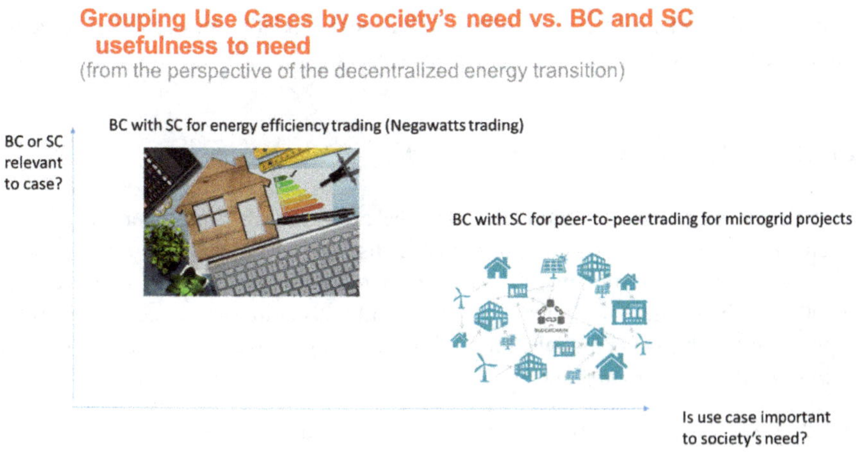

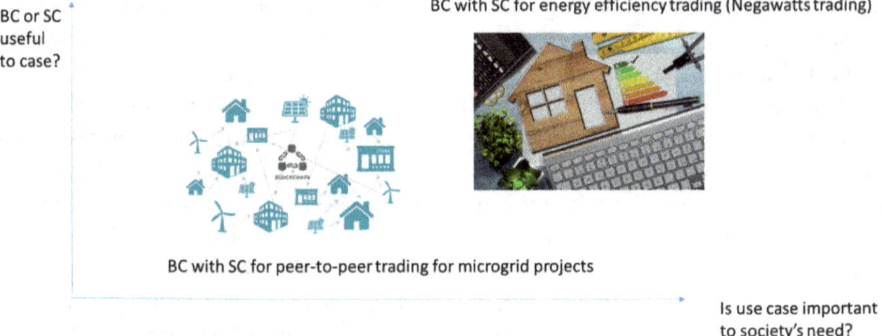

Fig. 10.11 Comparing chosen use cases under the perspective of a decentralized energy transition

Fig. 10.12 Comparing chosen use cases under the perspective of energy efficiency as key to the energy transition

energy efficiency trading market, even though regulations must follow to support such a strategy.

10.4.4 The Business Models for the Use Cases

Although we used a simplified version of the Business Model Canvas, we elaborated a complete business model canvas with the (obvious) missing elements. We also elaborated the links between the different business model parties in terms of value exchanged and the corresponding revenue streams.

10.4.4.1 The "Negawatt" Business Model

This business model allows for tokens to be created each time a certain amount of energy savings is produced and traceable, using smart meters, or other sensors. When enough tokens are generated (for example 20 tokens) then the tokens can be spent on an investment in Renewable Energy infrastructure (approved by new rules supporting this business model). Such a business model would assume that there is a *supporting policy framework*:

(a) a policy that requires energy savings to be generated by at least large consumers (and perhaps also smaller consumers like residential building owners) and
(b) a policy that provides the eligibility for tokens to be spent on renewable energy projects even if this is a voluntary policy invented by an industry group.

This means, for example, that the tokens generated and sold can only be used for certain types of renewable energy projects that are considered favorable for the region or considered to provide higher value than others which are happening anyway. An example is solar or wind energy projects versus hydro energy projects. The rules must be defined to ensure a high value for the market created (not to water down the value and create a market used only for green-washing purposes).

Another point one can add is the ability for entities to save or bank their tokens for future use, in the case that they are concerned about not meeting future energy saving targets themselves or have concerns about the price or value of tokens rising. This latter case would only be relevant if there was a possibility to trade energy savings generated on another market (like the white certificate market) and if energy savings generated could be banked and used in the future—allowing energy consumers to meet their future energy reduction targets with energy savings produced in the past (or allowing them to meet their current targets with energy savings produced by other entities).

Another option is that an *Energy Services COmpany* (ESCO) could sell energy savings (tokens) to the government allowing for some tokens to be buried, so to speak, and not returned to the market, if the government esteems that energy savings must be pushed forward even further. This purchase by the government could be made possible by using funds gathered from taxes (for example taxes on fossil fuel sales). This value network is illustrated in Fig. 10.13 and the full business model is depicted in Fig. 10.14.

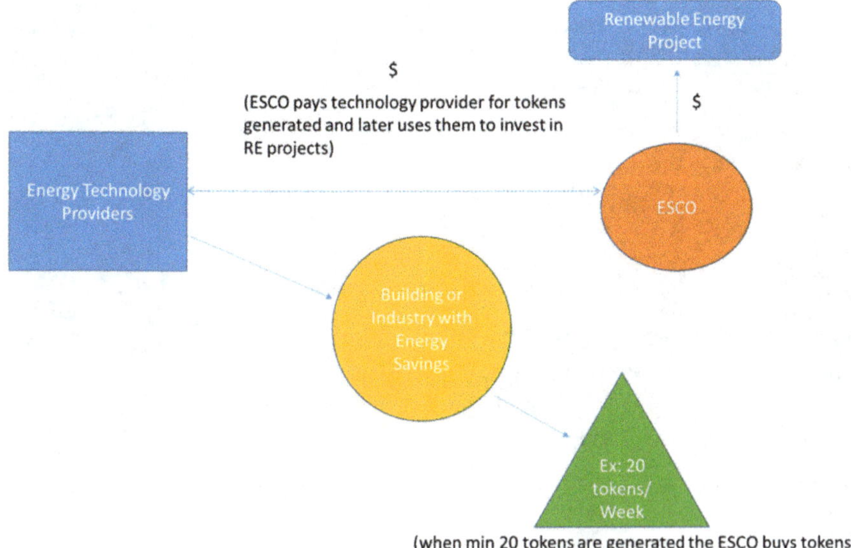

Fig. 10.13 Revenue streams and token relationships in the Negawatt business model

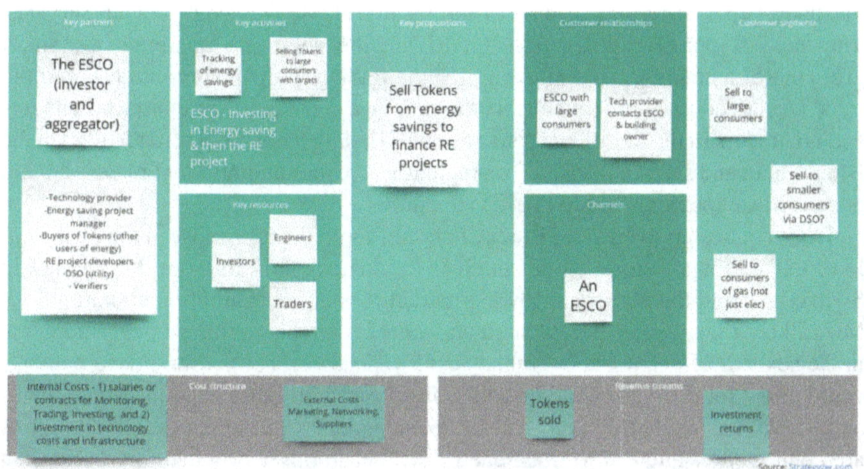

Fig. 10.14 Complete Negawatt business model using the traditional business model canvas

10.4.5 *Microgrid Business Model (Elaborated After the Workshop)*

The microgrid business model has not changed significantly after the workshop results. In the slightly revised business model (Fig. 10.15), we envision a P2P trading platform as the main channel for trading renewable energy generated, and

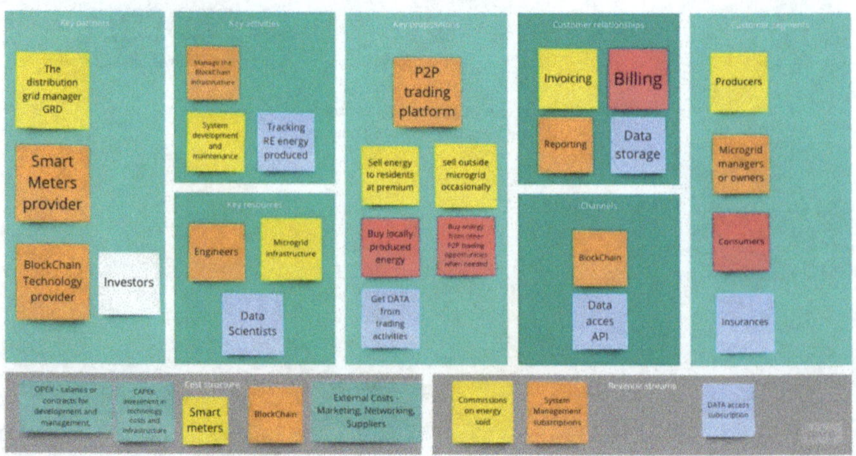

Fig. 10.15 Revised microgrid P2P trading business model using the traditional business model canvas. *Source* Strategyzer.com

the attributes it comes with. In addition, the trading of ancillary benefits (e.g., from aggregated data sold to insurance companies) could be envisioned as a second means to generate further revenue if the local grid has the need for such services. The main actors of this business model are the engineers operating the smart grid control technology and allowing for the trading with smart contracts, and the traders/smart contract proponents. By "traders" we mean more the contract managers and not traders in the traditional sense because trading happens automatically via the BlockChain and smart contracts. The regulators are also an important player, but not necessarily, depending on the market situation.

It has not been proven yet if BlockChain can actually make sense in the context of local P2P trading of energy (for example in the context of the microgrid (RCP) law in Switzerland and the potential growing market for auto-consumption communities). Most likely the value of BlockChain and smart contracts in this context will come in the next 10–15 years, but not in the next 5 years. Current microgrid developments could allow for P2P distributed trading of energy with other existing technologies and BlockChain will most likely add sufficient enough value once several microgrids are connected or allowed to trade energy with each other, or when microgrids can trade their energy with more than one local grid operator. More insights on this can be found in the literature reviews conducted in Bürer et al. [14, 15].

10.4.6 Discussion of the Results

From our current analysis of the workshop results, we can say that two use cases are at least seen by participants to provide potential value to society and to be clearly areas where BlockChain (or smart contracting) can really add value to the given goals outlined for each of the contexts provided.

Among both choices, we can also compare them in order to prioritize even further. But this depends on what we believe is more important. We could ask what is more valuable – whether the blockchain component is really necessary to unlock a critical underlying challenge like energy efficiency in buildings and industry for example, or if the scenario of distributed generation should be supported and therefore blockchain and smart contracts explored further even if current benefits from existing demonstrations have shown limited benefits on a local scale. Another test is to consider the difference between blockchain and smart contracts because smart contracts do not necessarily have to use blockchain technologies to function well. In other words, is it more important to focus on an area where BlockChain can provide more value to society or is it more relevant to focus on where smart contracting can offer more value.

Our assessment shows that maybe not BlockChain technology specifically, but smart contracting in general can provide great value to both contexts. If smart contracting can provide value to both energy efficiency innovations like our Negawatt application, as well as to local renewable energy markets, and blockchain adds more value to the Negawatt application than it does in the case of the microgrid case, we can assume that blockchain technologies are not necessarily the winning digitalization solution but that smart contracting is (sometimes enabled by blockchain technologies, and sometimes not).

Furthermore, if the view is that energy efficiency improvements are the first step toward an effective energy transition, then it makes full sense to explore further the Negawatt application in a next stage of exploration and business model conception, testing, and validation. This is also true because very little literature exists already on this potential use case. Furthermore, as we continued to develop the model behind the scenes (after the workshop) it was clear that the Negawatt concept could be combined also with renewable energy goals and therefore offer value on both fronts.

To conclude, both applications have value for society and must be further developed and tested in the future research work. This project helped to spur forward the knowledge in this area among researchers in academic institutions as well as the private sector and public sector stakeholders who participated in the workshop.

10.5 Current Trends and Future Scenarios

In this last section, we will present some new trends and future scenarios where BlockChain can enable further innovation in the area of Energy Transition.

10.5.1 P2P Distributed Energy Trading for Residential Microgrids

In Switzerland, Swissgrid is the TSO, that is, the company which owns and operates the national high-voltage transmission network. At the low-voltage level, 630 distribution system operators (DSOs) are in charge with distributing electricity to the consumers via the low voltage network. These DSOs are often very small companies which supply electricity for just one single commune. Only 30% of them produce also electricity on their own. The Quartierstrom P2P DET pilot project uses the utility grid operated by the local DOS, the Wasser-und ElektrizitätswerkWalenstadt (WEW). Because of that, the first operational P2P DET systems is based on a partnership with the existing local utility grid. Therefore, we consider the DSO as a relevant actor from a business model and value proposition perspective. We also consider the P2P Platform operator, who is in charge with the "virtual grid" and provides significant added-value services. The interconnection between elements of the virtual and physical grid are depicted in Fig. 10.16.

Quartierstrom is a pilot project led by the ETHZ and the University of Saint Gall. In the region of Walenstadt (SG), 37 households (27 prosumers and 10 pure consumers) are participating to a local electricity trading market. The P2P platform allows the project participants to buy and sell solar energy generated by PV modules directly between them, without intermediaries such as the utility grid or any other trusted third party [2, 10].

The automatic energy trading is made possible by the BlockChain platform based on Ethereum. Prices are set following a double auction mechanism where both

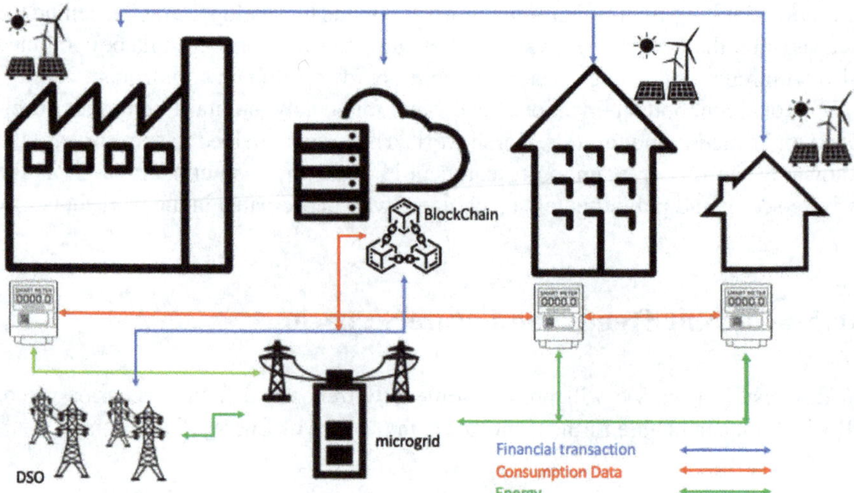

Fig. 10.16 Virtual and physical microgrids enabling P2P-DET applications

prosumers and consumers indicate the amount and the price at which they are willing to sell/buy locally produced energy.

All bids are transmitted by smart meters and registered in an order book and bids with lower selling prices and higher purchase prices are prioritized. The double auction algorithm is run iteratively every 15 min [10]. The discriminative pricing method is applied to each trade, which means that the final price is the mean between the highest buyer's price and the lowest seller's price [2]. Each transaction is implemented as a smart contract on the BlockChain [10].

The local grid operator and electricity supplier, the Water and Electricity Works Walenstadt (WEW), provides its distribution grid for the pilot test and collect grid usage tariffs [10]. In addition, the WEW contributes balancing the electricity supply and demand in buying surplus solar energy and supplying energy when the local demand exceeds the local supply [2]. Under the current Swiss legislation, private PV module owners have no choice but to sell solar energy to the public electricity supplier. However, given the experimental purpose of the project, arrangements were made to allow local producers to sell solar energy directly to local consumers [2].

10.5.2 *Bridging Industrial and Residential Microgrids Through Electric Vehicles*

The ultimate goal of microgrids is the full independence from the main grid, therefore supplying energy to members of the microgrid with local (possibly renewable) energy sources, without the need of balancing the energy supply and demand from external sources.

In many European countries or in big cities, people tend to live in multi-family or condominium housing structures, contrary to the typical individual house pattern of rural, semi-rural or peri-urban regions, not to mention the typical North American dispersed living landscape. This socio-demographic characteristic indeed has very profound consequences for the future penetration of *electric vehicles* (EV), due to the fact that people living in multi-family houses in many case do not have access to an individual parking place for their vehicle, that is usually parked on public domain or on other non-electrified spaces; correspondingly, that portion of the population cannot rely on an individual or even shared power outlet that could be reliably used to charge a private vehicle. It is thus easily understandable that switching to EVs is very negatively impacted by this all-too-common situation linked to past urban planning and architectural choices.

Hence, finding alternative charging solutions for dwellers without access to adequate power supply outside of ultra-fast charging stations—that can be assumed to remain significantly more expensive and less distributed on a given territory at least on the medium-term, represents a crucial endeavor. Concomitantly, it is certainly desirable that the power supplied to EVs is generated by renewable energy sources, typically by PV installations. This represents an ulterior "social handicap" for people who do not own an individual house or have access to a connected parking space, since they usually cannot install PV modules on the (shared) housing space they occupy.

In order to remove this important socio-technical obstacle to EV penetration, especially in dense urban areas, the concept proposed here suggests considering either the workplace or large commercial buildings—such as multi-functional malls or even multi-sport centers—as alternative EV charging spots. Indeed, industrial or service buildings—for example, large factories or administrative centers—usually have both large roof surfaces that could be used to install important PV capacities and broad either internal or external parking spaces reserved for their employees. Indeed, a paradigm shift (for slow and daily charging) is suggested in which workplace or shopping centers become main, standard EV charging spaces, instead of each one's residence.

The basic cornerstone of such a system would be the owner(s) of the requested surfaces for the PV modules. Key here is the scale economy that is expected by purchasing a large number of solar panels and the corresponding power control infrastructure. The CAPEX of this large implementation is expected to be significantly lower than the sum of individual, territorially decentralized PV modules investments. Therefore, the production price of the renewable power lies at significantly lower levels by working hypothesis.

In order to ensure the financial viability of the novel EV charging framework, two types of (smart) contracts between PV plant owners and customers—employees or customers of a given commercial mall—can be built:

(a) Energy-based contracting: The owner sells the electric power for EV charging at a price slightly higher than generation (including charges to cover CAPEX amortization, OPEX and investment for EV stationary charging infrastructure) for a fixed time period, with priority given. Business model must be designed in such a way that power is significantly cheaper than in standard charging stations and to cover expenses for grid imports in case of insufficient production.

(b) Surface-based contracting: The owner rents a given PV surface to employees or regular customers and the production is put at their disposal for EV charging. Rent must include charges for CAPEX amortization of charging infrastructure. In case the difference between production and consumption for EV charging is positive—on a time-basis to be defined contractually as well—, the owner can sell the overproduction to the local grid (and thus increase profitability of the scheme).

The proposed framework presents an additional significant advantage, that would bring interesting inputs for grids in residential areas and even for the future microgrids. Indeed, the framework is primarily intended for people who do not have access to adequate power supply at home for EV charging and their access should be prioritized. However, individual homeowners could also form a second circle of customers and, in fact, use the battery of their EVs, that they have charged on their workplace or at the commercial mall, as the battery for their houses (alone or in complementarity with a stationary storage). Hence, within the proposed paradigm change, EV batteries would/could provide support to the low-voltage distribution grids—for example, for covering the evening demand peak—or even support local future microgrids (based on separate contracts).

Since majority of the people work during daylight hours, it is indeed during work time that the probability is higher than sun shines and can be used to charge EVs; hence, our proposal to foster charging at workplace or at commercial sites, in the spirit of an EV charging shift adapted to the particularity of PV generation.

The role of BlockChain and Smart Contracts technology is fundamental in this application because it will make possible the above-mentioned contracting policy as well as the revenue sharing and the payments for EV charges. Hence, several microgrids can be dynamically connected through EVs which would efficiently store energy that can be used/charged just-in-time and just-in-place.

10.6   Conclusions

In this chapter, we have considered the role of BlockChain and Smart Contracts technology from a business-enabling perspective and provided a framework and a methodology for assessing the relevance of that technology in the context of the Energy Transition.

We also considered the current constraints for adoption of new business models enabled by those technologies as well as related issues and potential risks.

We have presented the methodology for designing new business models with a multi-stakeholder approach by using the Business Model Canvas framework. Two cases have been presented and discussed together with their implications from societal, environmental and technology perspectives.

One Swiss implementation of the Peer-to-Peer Distributed Energy Trading system has been presented and a new solution for its issue has been proposed for dealing with supply-demand imbalance.

In conclusion, we believe that BlockChain technology may contribute substantially in transforming the energy sector. However, the various limitations highlighted in this paper (grid reliability and security issues, energy consumption issues, regulatory risk, as well as its inherent levels of technological uncertainty for firms) need to be properly addressed.

The key reason, however, is related to the complexity of managing properly the grid (e.g., grid reliability and stability in heavily decentralized power exchange configurations), and this aspect linked to the complexity and risks involved in the application of complicated smart contracts to an already complex sector. This is at least our viewpoint with regard to the short term, while in the long term, we may indeed see a more cloud-supported energy system itself financed in a completely different way than today. This itself (the financing element stimulated by BlockChain-based fintech innovations) could lead at least toward a significant increase in the democratization of the ownership (and even the democratization of value creation) for future energy systems.

Today, what we think is more likely to occur is that new business models and new ways of thinking about energy services (and new value propositions) will be inspired or stimulated by the hype around BlockChain (whether it is indeed just hype, or not). In that way, one can see how BlockChain technology innovation could have indeed an impact on the energy transition even under a scenario where BlockChain applications do not undertake mainstream acceptance by energy sector stakeholders in the end.

The thinking around digital finance or new ways of financing renewable distributed energy systems, as inspired by BlockChain technology, may start to significantly change the way companies in the sector think about doing business and open their minds to new options with regard to financing models that support distributed generation and the systems needed to distribute energy in a heavily decentralized power exchange configuration.

Other than this, we have made the case that we still need the grid and that more work is needed to understand other implications of BlockChain in the energy industry, including how much energy BlockChain applications will consume if the world moves actually toward a much more connected and even AI-facilitated and virtual management of energy services and value propositions.

For the short term, we also see that the application of BlockChain technologies to electric power exchange at the local or regional level will have to physically rely on the grid itself. On the other hand, we may soon witness the creation of links between industries and non-traditional players with traditional players in the industry and pseudo innovation ecosystems that did not exist before because of BlockChain.

References

1. Abdella, J., Shuaib, K.: Peer to peer distributed energy trading in smart grids: A survey. Energies **11**(6), 1560 (2018)
2. Ableitner, L., Meeuw, A., Schopfer, S., Tiefenbeck, V., Wortmann, F., Wörner, A.: Quartierstrom-Implementation of a Real-World Prosumer Centric Local Energy Market in Walenstadt. Switzerland. arXiv:1905.07242 (2019)
3. Ahl, A., Yarime, M., Tanaka, K., Sagawa, D.: Review of blockchain-based distributed energy: Implications for institutional development. Renew. Sustain. Energy Rev. **107**, 200–211 (2019)
4. Apajalahti, E.L., et al.: From demand side management (DSM) to energy efficiency services: a finnish case study. Energy Policy **81**, 76–85 (2015)
5. Baran, P.: On distributed communications. RAND Corp No. RM3420PR, Santa Monica, CA (1964). http://www.rand.org/pubs/research_memoranda/RM3420.html
6. Birkin, F., Polesie, T., Lewis, L.: A new business model for sustainable development: an exploratory study using the theory of constraints in nordic organizations, July 2009. Bus. Strategy Environ. **18**(5), 277–290 (2009)
7. Blyth, W., et al.: Low Carbon Jobs: The Evidence for Net Job Creation from Policy Support for Energy Efficiency and Renewable Energy. UKERC (2014)
8. Bocken, N.M., Short, S.W., Rana, P., Evans, S.: A literature and practice review to develop sustainable business model archetypes. J. Clean. Prod. **65**, 42–56 (2014)

9. Boons, F., Lüdeke-Freund, F.: Business models for sustainable innovation: state-of-the-art and steps towards a research agenda. J. Clean. Prod. **45**, 9–19 (2013)
10. Brenzikofer, A., Meeuw, A., Schopfer, S., Wörner, A., Dürr, C.: Quartierstrom: a decentralized local P2P energy market pilot on a self-governed blockchain. CIRED 2019 Conference (2019)
11. Brunner, et al.: Einführung in die Pinch-Methode; Handbuchfür die Analyse von kontinuier-lichenProzessen und Batch-Prozessen - ZweiteAuflage. BundesamtfürEnergie BFE (2017) (in German, available at http://www.bfe.admin.ch)
12. Budde Christensen, T., Wells, P., Cipcigan, L.: Can innovative business models overcome resistance to electric vehicles? Better place and battery electric cars in Denmark. Energy Policy **48**, 498–505 (2012)
13. Burger, S.P., Luke, M.: Business Models for Distributed Energy Resources: A Review and Empirical Analysis, MIT Energy Initiative Working Paper (2016). https://energy.mit.edu/wp-content/uploads/2016/04/MITEI-WP-2016-02.pdf
14. Bürer, M.J., De Laparrent, M., Pallotta, V., Capezzali, M., Carpita, M.: Use cases for blockchain in the energy industry: A critical review. In: Proceedings of ICE/IEEE ITMC Conference (2018)
15. Bürer, M.J., De Laparrent, M., Pallotta, V., Capezzali, M., Carpita, M.: Use cases for blockchain in the energy industry opportunities of emerging business models and related risks. Comput. Ind. Eng. **137**, 106002 (2019)
16. Deloitte: Blockchain Risk Management, Risk Functions Need to Play an Active Role in Shaping Blockchain Strategy (2017)
17. Depsys: (2020). GridEye:https://www.depsys.ch/researchprojects/grideye/
18. EU 2017: accessible via https://ec.europa.eu/commission/second-report-state-energy-uni on_en
19. Geels, et al.: Nat. Clim. Change (2016). https://doi.org/10.1038/NCLIMATE2980
20. Geissdoerfer, M., Vladimirova, D., Evans, S.: Sustainable business model innovation: a review. J. Clean. Prod. **198**, 401–416 (2018)
21. Gochenour, P.H.: Distributed communities and nodal subjects. New Media Soc. **8**(1), 33–51 (2006)
22. Gralla, et al.: Renew. Sustain. Energy Rev. **70**(2017), 1251–1265 (2017)
23. Gui, E.M., MacGill, I.: Typology of future clean energy communities: an exploratory structure, opportunities, and challenges. Energy Res. Soc. Sci. **35**, 94–107 (2018)
24. Hall, S., Foxon, T.: Values in the smart grid: the co-evolving political economy of smart distribution. Energy Policy **74**, 600–609 (2014)
25. Hall, S., Roelich, K.: Local Electricity Supply: Opportunities, Archetypes and Outcomes. Ibuild/RTP Independent Report (2015)
26. Hannon, M.J., et al.: The co-evolutionary relationship between energy service companies and the UK energy system: implications for a low-carbon transition. Energy Policy **61**, 1031–1045 (2013)
27. Higginson, M, et al.: The promise of Blockchain. McKinsey.com (2017)
28. Hultman, et al.: Energy Policy **40**(2012), 131–146 (2012)
29. IEA: Capturing the Multiple Benefits of Energy Efficiency (2014)
30. Infosys: https://www.infosys.com/Oracle/white-papers/Documents/carbon-supply-chain-blo ckchain-technology.pdf (2018)
31. Johnson, M.W., Suskewicz, J.: How to jump-start the clean tech economy. Harvard Bus. Rev. **87**(11), 52–60 (2009)
32. Johnson, M.W., et al.: Reinventing your business model. Harvard Business Review (2008)
33. Joyce, A., Paquin, R.L.: The triple layered business model canvas: a tool to design more sustainable business models. J. Clean. Prod. **135**, 1474–1486 (2016)
34. Markides, C., Geroski, P.A.: Racing to be second. Business Strategy Review 25. London Business School (2004). https://onlinelibrary.wiley.com/doi/epdf/10.1111/j.0955-6419.2004. 00336.x
35. Martin, C.: How Blockchain is Threatening to Kill the Traditional Utility. BNEF (2018)
36. Mengelkamp, E., Gärttner, J., Rock, K., Kessler, S., Orsini, L., Weinhardt, C.: Designing microgrid energy markets: a case study: the Brooklyn microgrid. Appl. Energy **210**, 870–880 (2018)

37. Michaelides, et al.: Nucl. Eng. Des. **366**(2020), 110742 (2020)
38. Mont, O., Dalhammar, C., Jacobsson, N.: A new business model for baby prams based on leasing and product remanufacturing. J. Cleaner Prod. **14**(17), 1509–1518 (2006)
39. Osterwalder, A., Pigneur, Y.: Business Model Generation: A Handbook for Visionaries, Game Changers, and Challengers. Wiley (2010)
40. Pietzker, et al.: Energy **64**(2014), 95–108 (2014)
41. Pouttu, A., Haapola, J., Ahokangas, P., Xu, Y., Kopsakangas-Savolainen, M., Porras, E., et al.: P2P model for distributed energy trading, grid control and ICT for local smart grids. In: 2017 European Conference on Networks and Communications (EuCNC), pp. 1–6. IEEE (2017)
42. Provance, M., Donnelly, R., Carayannis, E.: Institutional influences on business model choice by new ventures in the microgenerated energy industry. Energy Policy **39**, 5630–5637 (2011)
43. Roelich, K., et al.: Towards Resource-Efficient and Service-Oriented Integrated Infrastructure Operation. Technol. Forecast. Soc. Change **92**, 40–52 (2015)
44. Strupeit, L., Palm, A.: Overcoming barriers to renewable energy diffusion: business models for customer-sited solar photovoltaics in Japan, Germany and the United States. J. Clean. Prod. **123**, 124–136 (2016)
45. Schneider, J., Blostein, A., Brian, L., Kent, S., Groer, I., Beardsley, E.: Blockchain: putting theory into practice. Profiles in Innovation Report, The Goldman Sachs Group (2016)
46. Stöcker, C.: BlockCharge—EV Charging via the Ethereum BlockChain. [Online] YouTube. Available: https://www.youtube.com/watch?v=0A0LqJ9oYNg. Accessed 18 Aug 2016
47. Swan M.: Blockchain: Blueprint for a New Economy. Newton, MA, USA: O'Reilly Media (2015)
48. Unruh, G.C.: Escaping carbon lock-in. Energy Policy **30**(4), 317–325 (2002)
49. WEF: Can blockchain help us to address the world's energy issues? By Jon Creyts and Ana Trbovich, an article part of the World Economic Forum Annual Meeting (2018)
50. WEO: Blockchain—an opportunity for energy producers and consumers? IEA, World Energy Outlook (2015). https://www.pwc.com/gx/en/industries/assets/pwc-blockchain-opportunity-for-energy-producers-and-consumers.pdf
51. Wainstein, M.E., Bumpus, A.G.: Business models as drivers of the low carbon power system transition: a multi-level perspective. J. Clean. Prod. **126**, 572–585 (2016)
52. Wright, A., De Filippi, P.: Decentralized Blockchain Technology and the Rise of Lex Cryptographia (March 10, 2015). Available at SSRN: https://www.ssrn.com/abstract=2580664 or https://doi.org/10.2139/ssrn.2580664
53. Zhang, C., Wu, J., Long, C., Cheng, M.: Review of existing peer-to-peer energy trading projects. Energy Proc. **105**, 2563–2568 (2017)

Chapter 11
The Feasibility and Significance of Employing Blockchain-Based Identity Solutions in Health Care

Peng Zhang and Tsung-Ting Kuo

Abstract The wide adoption of wireless communication and mobile devices has facilitated the development of numerous applications to provide citizens with convenient access to health-related tracking and management services. Most of those services require the storage of some personal data and therefore resort to common user authentication practice (e.g., using a username and password combination) to ensure data is delivered to the appropriate party. As a result, users often find themselves having to maintain or memorize many combinations of accounts and their associated login credentials during their interaction with different services throughout the lifespan. Given the advancement of blockchain and distributed ledger technologies, a wealth of services in various domains including health care has explored the feasibility of migrating existing centralized services to such decentralized infrastructures. Because of this exploration, traditionally centralized authentication approaches managed by one party can no longer support the need of onboarding users, managing, and monitoring user activities and transactions in a decentralized manner. A community of researchers has hence been formed to study blockchain-based identity solutions, such as decentralized identities and self-sovereign identities, that would allow users to have a more common way to identify themselves when accessing a plethora of services. The main goal of these identity methods is to eliminate the need of requiring users to maintain multiple identifiers or online credentials as each individual has only one identity that truly represents themselves. These identities would be established and secured by cryptographic principles such that they still preserve at least the same security and privacy levels as their centralized counterparts. In this chapter, we first present a systematic overview of the underlying motivations and principles of blockchain-based identities to provide the audience with a basic understanding of how such identities operate and the pressing need to incorporate them. We will also introduce two of the popular blockchain-based identity frameworks currently adopted in decentralized applications. We then discuss the potential

P. Zhang (✉) · T.-T. Kuo
Belmont University, Nashville, TN, USA
e-mail: peng.zhang@vanderbilt.edu

T.-T. Kuo
University of California San Diego, San Diego, CA, USA

applications of these identities and their feasibility using the health care domain as a case study to hopefully inspire our readers with ideas that can be further investigated as research solutions in the health care or other domains. Lastly, we will conclude the chapter with additional discussions on the practicality of blockchain-based identities and the potential caveats or limitations associated. This chapter will serve as a cornerstone for healthcare executives, informaticians, and security/privacy experts to further investigate and make infrastructural decisions.

11.1 Introduction

An identity is a vital part that identifies and describes an individual. It is composed of a wide variety of information, such as first and last names, date of birth, home address, telephone number, family members, biometrics, whose combination is uniquely possessed by an individual. In reality, only a basic portion of the identity is needed as an identifier for a person to open a new bank account, purchase a home, receive medical service, pick up a prescription, or accomplish other numerous tasks with. An identifier generally takes place in the form of identification documents, such as passport, state government issued photograph ID, and birth certificate, for official or in-person identity verifications, or an email/password combination or mobile device authentication for online verifications. As communications become much more convenient and rely less on in-person contact thanks to the ubiquity of the Internet and smart personal devices today, fragments of individual identities have been distributed to a large number of agencies during the online use and submissions of various types of identifiers. Many owners of those identities are unaware of the extensive presence of their information because of scattered intermediary data collectors [35]. For users, protecting the privacy and safety of involved identifiers is thus equally critical as being able to verify identities, to safeguard against abuse and attacks in the modern era.

Fragmented identities pose many challenges in real-world situations where identity verifications play a significant role. Healthcare providers, as one example, regularly collect identifiable information, medical data, drug allergies, etc., of patients prior to offering medical service. However, due to the lack of standardized representation of patient identities and inconsistencies in the way such data is captured across different health IT systems, a patient may be requested to provide the same exact information each time at a new or regularly visited physician's office. Similarly, often during clinical trial recruitment, volunteers interested in participating in multiple trials repeatedly answer the standard set of questions regarding their identification and demographic information [36]. Health care is certainly not the only industry where users witness and engage in the same data collection process that has long been the common practice in day-to-day functions. Regardless, it is inefficient, error-prone, and creates fatigue for identity owners and verifiers due to the repetitiveness and complexity in information presentation and verification.

The root cause of the fragmented identities followed by repeated data collection is the excessive degree of centralization that exists today, which locks the information in the siloed system where it is obtained. In recent years, there has been growing interest in the construction of a new, decentralized identity model to minimize the pain points of centralization. The primary goal is to return the ownership, management, and access control of identities and identifiers alike to the original, individual identity hosts. The concept of decentralized identities exploits the decentralized and tamper-resistant nature of the emerging blockchain technologies to offer a much more structured and permanent identity for each person or entity that needs to be identified. In this chapter, we first present a brief evolution of identities in the digital age, leading to discussions of the underlying motivations and principles of blockchain-based identities. We then introduce two existing decentralized identity frameworks, uPort and Sovrin, followed by discussions of the potential applications, and limitations of these identities in the health care domain to inspire future research explorations of decentralized identity models.

11.2   A Brief History of Identities in the Digital Age

This section presents a short history of user identity management (IdM) models in the Internet era, focusing on three phases of centralized digital identities: fully centralized identity management [1], federated identity management [6], and user-centric identity management [14]. Through the development of these models, the limitations of centralized management are exposed, and the demand to move beyond centralized models becomes clear.

11.2.1   Fully Centralized Identity Management

The advent of the Internet has revolutionized the world by virtually connecting everyone digitally and sparking immeasurable opportunities that enabled much more speedy, convenient, and easily accessible communications and services. The Internet has empowered its users to engage in a broad range of activities of their choice, such as publishing and browsing web pages, conducting business or performing job duties, managing financial accounts, and making medical appointments, all at the comfort of their homes. With the connectivity and accessibility to services across different sectors, it has become apparent that the way modern citizens are building their online presence is much more diversified than that prior to the Internet era. At the same time, users' identities are being progressively splitted and replicated.

In this digital age, it is commonplace for an Internet user to have multiple email accounts, one for personal use, another for business purposes, and perhaps one more for subscription management. Any one of the accounts may be used to set up the user's social or business profile, online banking or health services account, and so on. Each service that features personalized experience typically creates a profile for the user with a login credential, e.g., a combination of username and password. This common model marks the first stage of identity development in the digital age–**fully centralized identity management** [1].

Although full centralized IdM has matured over the years, its widespread use has exposed several key limitations. First, the vast majority of identities captured with this type of system today lack standardization, which makes the verification of an identity heavily dependent on the identity manager, that is, the product owner or service provider. The most popular verification methods offered by online services today include a username and password combination, an email address and password, a phone number and password, or a phone number and a phone verification code requested and entered during login. As a rising number of services becomes available, an average user is expected to remember or keep track of a large amount of login information to access their desired services. Second, if a user forgets or loses any part of the login information to a particular service, they have to go through a process to retrieve or reset that information. However, the exact retrieval process varies from service to service. Imaginably, some services with strict account security protections for necessary reasons (e.g., personal banking accounts or patient billing accounts) would require more identifications to be verified than others (e.g., retail services). Additionally, stricter login retrieval may require identification documents, which could also change over time (e.g., as name or address changes, or lost documents are replaced), and these situations would further complicate or delay the account retrieval process. Therefore, fully centralized IdM can lead to user fatigue and frustration, and, in the worst case, a user may permanently lose access to their account or activity history because, in reality, it is each service provider instead of the user who owns the user's identity information.

11.2.2 Federated Identity Management

Moving forward, many businesses have realized the shortcomings of fully centralized IdM and are working to ease the "one account per service" burden and enhance user experience. In recent years, larger-scale service providers, with Google and Facebook being the most popular, have implemented the OAuth 2.0 standard protocol [12], which allows their users to provide authenticated access to other online applications using their existing accounts with these services [7]. In other words, services like Google and Facebook became the central identity providers for both users and other client applications with whom they have established trust relationships. This identity model shifts the individual centralization to a limited number of services and is known as **federated identity management** [6].

Federated IdM mitigates the issues of user fatigue/frustration, while it raises another critical disadvantage: under this model of user identity authentication, it is natural for user information to be monopolized by the "Internet Giants." On the one hand, these companies indubitably uphold stringent data security protection measures, are equipped with extremely high-performance databases, and have already maintained a substantial and stable user population. On the other hand, they are the most attractive targets to attackers because they possess an enormous amount of users' sensitive, personal information and, in addition to that, users' authentication history with other services. Facebook's recent data breach incidents [22, 34] are one instance of how a federated identity model that remains centralized in nature does not reduce users' susceptibility against information leakage. For this reason, it is uncommon today for hospital patient portals to adopt federated identity management.

11.2.3 User-Centric Identity Management

In 2008, Kim Cameron, Reinhard Posch, and Kai Rannenberg published a proposal [4] on user-centric IdM to improve upon centralized and federated models. Based on observations of the incompatibility across various IdM systems at the time, they introduced a framework involving various delegated roles with standardized components that allow users to freely control the flow of their information. The authentication process starts with a user-initiated claim of an identity ownership that then gets validated by a trusted administrative party selected by the user. The result of the verification is consequently relayed to the actual service that consumes this claim and provides the requested service to the user. This framework still implied some dependencies on centralized/federated roles to carry out critical tasks such as storing information, so the risk of potential privacy breach remains. It also has not been implemented fully to prove its practical use. Nevertheless, the idea was a step closer to granting the true identity owners some control of their information.

11.3 The Main Ideas Behind Decentralized Identity Management

To mitigate the concerns of having a central point of attack presented by centralized IdM models, the idea of adopting decentralized IdM, which removes the dependency of central identity providers, has emerged recently [7]. This section provides an overview of the main ideas, including decentralized identifiers and a method known as zero-knowledge proof, that set decentralized IdM apart from its centralized predecessors.

11.3.1 Decentralized Identifier Based on Blockchain

As communities began to recognize the need for digital IdM to move beyond centralized control, blockchain technologies that promote decentralization started garnering attention due to its successful use cases in cryptocurrencies. The key attraction of cryptocurrencies lies in their support of so-called trustless financial transactions between users that require no intermediaries, which means that only the parties directly affected are involved in each transaction. As the cryptocurrency pioneer, Bitcoin [19] has sustained to date without any fundamental failure since its inception. Another powerful decentralized computing ecosystem, Ethereum [33] also has fostered many decentralized applications beyond cryptocurrency exchanges. With the increasing capabilities, blockchain technologies have also cultivated the idea of **decentralized identity management** (DecIdM).

DecIdM system builds atop decentralized blockchain technologies to remove much of the dependency on single third-party mediators between users and their requested service providers. The disintermediation facilitates identity model designs with a focus on returning complete access control and management rights to users. Users can then truly become the holders and managers of their own decentralized identifiers (DIDs) [29] and are free to distribute their DIDs to other parties who can directly consume this data without having to create new identities. A DID is quite similar to an identifier of a cryptocurrency user, which is similar to an easily verifiable and secure public-private key pair [19]. In the naive form, DIDs have a public portion resembling a digital address that is to be openly shared with others and a private portion that is a secret key to be kept to the DID owner. In order to ensure that a DID exists and is unique, the public address of that DID must be registered in the underlying blockchain framework.

When a DID is first generated, it is only supplied with an empty address with little meaning about an identity. It is a free choice of the DID holder to include any additional aspects of their profile as necessary, and all aspects can be accessed from a single location instead of being arbitrarily disseminated. These aspects represent information that can be verified by other parties in order to offer services or handle requests based on user agreements or regulations. Gender, age, nationality, occupation, and educational level are example aspects to be associated with a DID.

In order to link any kind of a digital identity to its corresponding physical identity, a portion of its aspects requires an endorsement from designated authorities offline. The remaining aspects of that DID must be validated from other parties who can be held accountable for their authenticity. In other words, aspects of a DID that involve official reputation, such as legal name and date of birth, must be directly attested by authoritative agencies, while other aspects can be verified by other DID holders who have been directly or indirectly endorsed by authorities. As proof of verification, a digital certificate is created and signed by the verifying DID holder. The certificate can then be shared or used to build a chain of related certificates with non-repudiation. Figure 11.1 shows the interactions between DID holders who have different roles in the physical world.

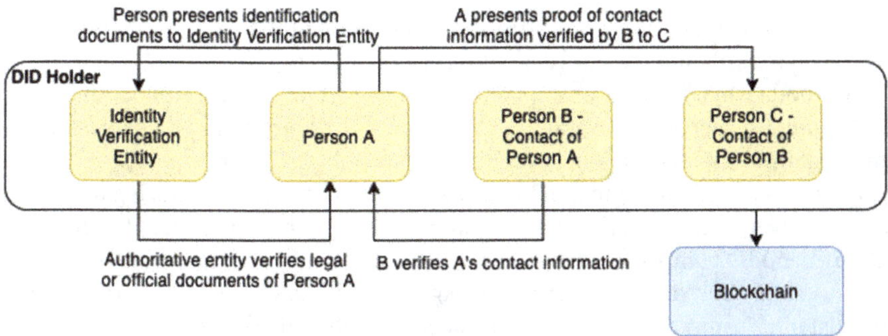

Fig. 11.1 Interactions between DID holders with different physical roles

11.3.2 Zero-Knowledge Proof

To preserve the privacy of personal information, specific aspects of a DID can be encapsulated by the DecIdM service through a method known as zero-knowledge proof (ZKP) [25]. In short, ZKP would allow a DID holder to prove an aspect of their identity without requiring any specific information of that aspect to be disclosed to other parties. One way ZKP can be implemented in DID systems is to delegate the verification tasks to a number of trusted authorities (e.g., a government office or a notary service), who each holds a respective DID. Any authority selected by the DID holder can provide a certificate of their claim along with a signature generated from the authority's DID and an expiration date of the certificate. The certificate and its expiration date will be recorded in the blockchain (equivalent to a record of some cryptocurrency exchange) and attached to the DID of the owner making the claim. The signature from the authority can be verified using cryptography, just like verifications of signatures in a cryptocurrency blockchain [20]. The DID owner can then share the certificate with any other user to prove the possession of a certain aspect of their identity. In doing so, both the claim initiator and the verifier of the claim are liable for the certification process in case of any dispute. Although ZKP might not be a required component, most DecIdM systems implement ZKP or other cryptographic techniques to protect user privacy.

Here is a simple example demonstrating how ZKP would work: suppose that DID holder Alice would like to establish an aspect of her identity as an insured patient by claiming that she currently receives healthcare insurance benefits from Company X in order to receive care. Alice can request Company X to serve as her certifying authority for the insurance claim. Alice may contact Company X offline to provide the required information (or other certificates already established as part of her DID). When the insurance agent completes the verification of the documents offline, they can issue a formal verification of Alice's current insurance enrollment by signing the certificate with an expiration. This record is logged in the blockchain transaction list and linked to Alice's DID. Alice is now able to share her DID with a new certificate showing her proof of insurance coverage to any healthcare provider she chooses, without the need to disclose personal identifiable information such as her social security number and date of birth.

As another hypothetical example, the US laws prohibit anyone under the age of 21 to purchase alcoholic beverages [31]. In order for stores to avoid charges of selling alcohol to minors, they may require an alcohol purchaser to present a state or government issued photograph ID to prove that they have reached the legal drinking age. However, relatively private information on the ID, including date of birth, home address, and ID number, is unnecessarily disclosed to the store associates–they only need to verify that the customer is over age 21. With ZKP implemented in a DID, the customer can prove that they are indeed over the drinking age limit via a DID certificate previously issued from their state government. In this situation, the customer no longer needs to present other irrelevant personal information as long as the signature signed by the certifier can in fact be validated against the public portion of their DID.

11.3.3 Decentralized Blockchains and the Internet

To a certain degree, decentralized blockchains powering DIDs mimic three characteristics of the Internet. First of all, no single user controls them, as they are built on a distributed rather than a centralized network. Second, everyone can access and use them, because data is replicated by an indefinite number of users and thus becomes available for everyone. Third, everyone can improve them by independently appending or updating the information, although removal is much more difficult to realize within blockchains. Blockchains' similarities to the readily available Internet today suggest their strong potential in successfully supporting decentralized identities as their own independent ecosystem. Combined with concepts and practices that have matured since the early days of the Internet, digital identities can be designed with standardization to maximize compatibility between different versions and/or implementations. Advanced technologies such as ZKP can also be integrated into the designs early on to provide an added privacy layer that meets the needs of health care as well as other applications.

11.4 The Technical Architecture of Decentralized Identity Management

In this section, we present an overview of the technical architecture of DecIdM and introduce two decentralized identity frameworks that have been implemented today.

11.4.1   Architecture Overview

Figure 11.2 presents a high-level architecture of a blockchain-based DecIdM system. The **Blockchain** component serves as a ledger storing verifiable and immutable transactions as they are appended to the **Blockchain**. Despite the different roles DID owners have in the physical world (as denoted in parentheses in each DID Holder box), the **DID Holder** component is agnostic about those roles. It simply represents an identity of a **Person** or an **Entity** alike and consists of different aspects of that identity. A DID Holder can make requests to the **DID Service** component, which is the glue of DecIDM that runs within the **Blockchain** (e.g., as a smart contract [3]) and is thereby decentralized. The requests can include generating a new DID, creating a claim of an aspect of a DID, distributing claim verification requests and results, and supporting verification queries of a DID claim. As each service request is processed, DID Service records related activities as Blockchain transactions and returns the response and result to the requesting DID Holder.

An example workflow of the DID verification process is shown in Fig. 11.3. To onboard the system, a Person (or an Entity) requests the DID Service to generate a new DID as their identity (step 1). Next, DID Service creates the new DID following some special template (discussed in the next subsection) and logs the public metadata of the identifier onto the Blockchain (step 2). It then returns the entire DID to the requestor without holding the private key (step 3a), and at this time, an initial link is established between the physical identity of Person and their DID (step 3b). As a new DID Holder, a Person can now proceed with adding a first aspect to their identity by making a subsequent claim to DID Service (step 4). The identity verification may be completed outside the DecIdM system directly between Person and any other DID Holder, such as Notary Service (step 5), who verifies some information about Person (step 6). Upon successful verification, Notary Service creates and forwards a certificate signed with its DID for the verified claim of Person to DID Service (step 7). DID Service records the certificate as a Blockchain transaction (step 8) and

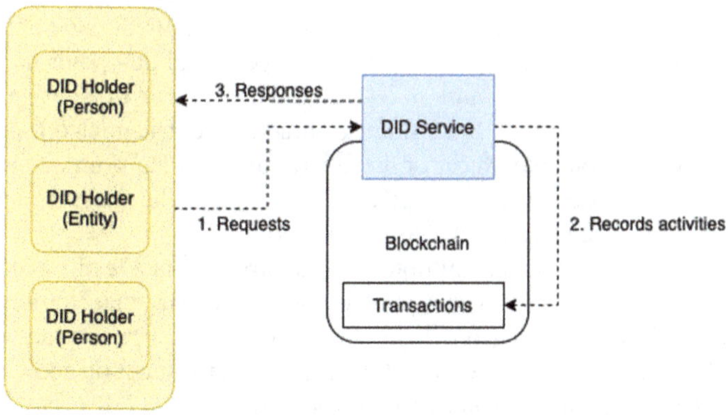

Fig. 11.2 A high-level architecture of blockchain-based DecIdM system

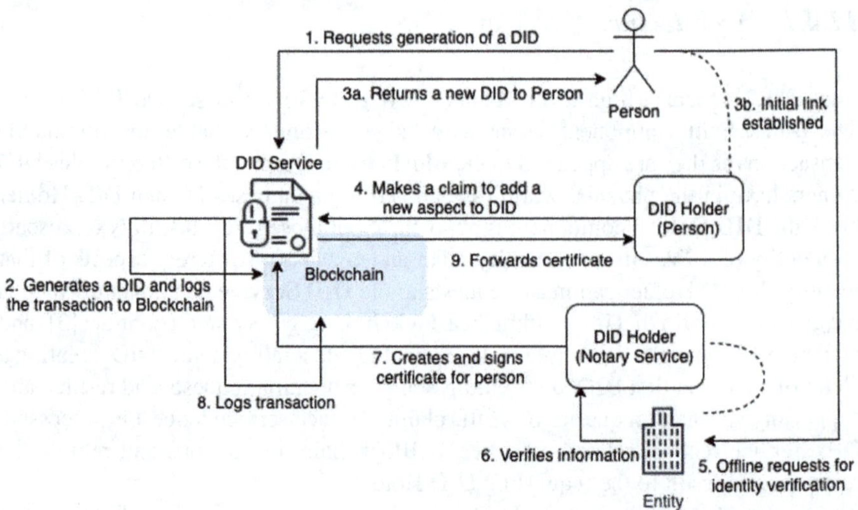

Fig. 11.3 A demonstration of the DID verification process

forwards the certificate to the DID Holder (Person), marking the successful addition of a first DID aspect. Two example DID frameworks will be introduced later in this chapter to illustrate implementation-specific processes.

11.4.2 Blockchain-Based DID Service

As the most important module of a DecIdM system, DID Service must (1) serve as a decentralized component to make sure that its service remains available to users and is not subject to any single point of failure as often experienced in centralized systems [26] and (2) handle all requests from DID Holders in a secure and privacy-preserving manner. To achieve these two goals, DID Service exists within a blockchain infrastructure. A form of DID Service would be one or more "smart contracts" [3], which contain program code that is guaranteed to be executed when predefined conditions are met. Each smart contract, once instantiated (i.e., the process of deploying a concrete object of the smart contract), has a unique address on the blockchain, and each access or method call to change the state variables of the smart contract is logged as a blockchain transaction.

A DID Service may delegate multiple smart contracts to tackle different subproblems. For instance, one contract can serve as a global registry that stores a mapping of all available DIDs and their on-chain "locations," i.e., smart contract addresses; whereas others handle new user onboarding, manage authentication and claims verification requests, implement protocols to replace lost or stolen DIDs, etc. The actual mechanisms of the contracts depend on the specific DID system implementation, of course. Some implementations may include protocols that can seamlessly handle requests to replace or remove any or all aspects of a DID (or, a user's entire association with a particular DID) so that user privacy can be protected. A decentralized DID service is a lot more challenging to implement compared to a centralized service because much more stringent computation and storage requirements apply when achieving decentralization.

11.4.3 Standardization of DIDs

Although decentralization removes many middlemen between users and their service providers, this feature alone is not sufficient to solve the predicament of users feeling overloaded with multiple identifiers among different services. Without any standardization, DID representations would continue to be implementation-specific. If DIDs generated by different systems are incompatible with one another, a limited number of services again will eventually monopolize the DID market. Fortunately, a community of experienced researcher teams led by Microsoft, IBM, MasterCard, etc., have developed open standards (e.g., W3C DID [29]) for services to adhere to when designing new DID models and their operations. The standards focus on a multi-level and multi-module DID structure to accommodate the varying volumes and types of information held by the identities. Figure 11.4 is an example from the W3C DID v1.0 standard showing the structure and a minimal set of attributes of a DID document [29].

```
{
  "@context": "https://www.w3.org/ns/did/v1",
  "id": "did:example:123456789abcdefghi",
  "authentication": [{
    // used to authenticate as did:...fghi
    "id": "did:example:123456789abcdefghi#keys-1",
    "type": "Ed25519VerificationKey2018",
    "controller": "did:example:123456789abcdefghi",
    "publicKeyBase58": "H3C2AVvLMv6gmMNam3uVAjZpfkcJCwDwnZn6z3wXmqPV"
  }],
  "service": [{
    // used to retrieve Verifiable Credentials associated with the DID
    "id":"did:example:123456789abcdefghi#vcs",
    "type": "VerifiableCredentialService",
    "serviceEndpoint": "https://example.com/vc/"
  }]
}
```

Fig. 11.4 An example from the W3C DID v1.0 standard document [29]

In the example, attribute "@context" specifies the type and version of the standards this document implements, and "id" represents the unique ID of the owner or subject this document refers to .[1]. The "authentication" field lists a set of verification methods the DID holder has authorized–in this case, the holder (indicated by the "controller" attribute) has only one method of authentication, which is a private key that pairs with the given public key (denoted by "publicKeyBase58") of type "Ed25519VerificationKey2018." Finally, the "service" list contains one URL (noted by "serviceEndpoint") at which a certificate (or a "Verifiable Credential" in the W3C standard) verifying some claim can be retrieved from "VerifiableCredentialService" service, indicated by attribute "type" with the DID specified in attribute "id".

11.4.4 Examples of DID Frameworks

11.4.4.1 UPort

uPort [17] is an open source DecIdM system built upon the public Ethereum blockchain via Ethereum's native smart contracts. It is created to allow users to host and develop their own identity, directly interact with other identity owners, and selectively distribute their identity to others in a secure manner with non-repudiation. Its identity management is achieved through four main components of the architecture: a mobile application, a controller contract, a proxy contract, and an identity registry contract. Figure 11.5 is a simplified diagram showing the relationships between the discussed components of the uPort system.

The uPort mobile application ((a) in Fig. 11.5) is decentralized and provides a user-friendly interface to handle the new identity generation and credential establishment [17]. The application forwards user requests to the appropriate smart contracts for processing, with each activity being recorded on the blockchain as a verifiable transaction. On start, the application generates a new identity for its user as a cryptographically linked key pair. The pair has a public key that is linked to other uPort contracts and a private key that is securely stored on the user's mobile device. Subsequently, the public key is used to deploy an instance of a uPort contract called "controller" ((b) in Fig. 11.5) that primarily holds the identity recovery and access control logic. The deployment of a controller then triggers the instantiation of the "proxy" contract ((c) in Fig. 11.5) that references the address of the controller and serves as the global identifier of the user, namely, uPort identifier. The proxy contract instance is a delegate that transacts with other smart contracts on behalf of the controller, and its functions can only be invoked by the controller upon the user's interaction with the mobile application. This mechanism introduces an additional layer that allows the user to recover their identity in the case of lost or stolen private key. During the private key recovery process, the user self-elects a group of trusted uPort identity owners (e.g., close contacts, financial institutions, etc.) who can attest

---

[1]The complete standards document can be found at https://www.w3.org/TR/did-core.

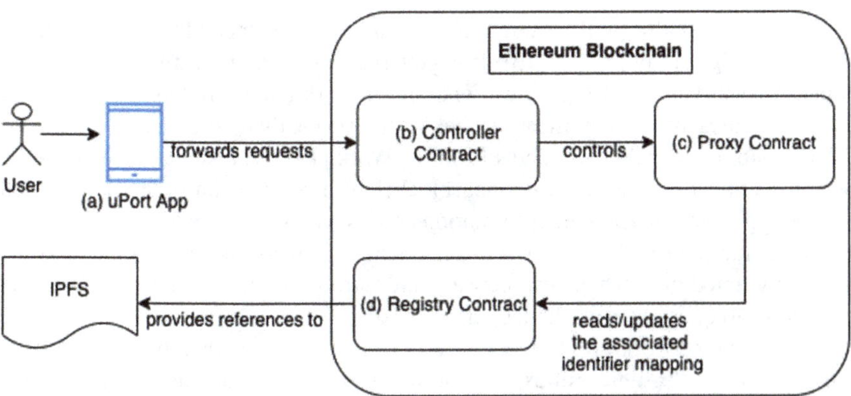

Fig. 11.5 A simplified diagram of the interactions between key components of the uPort IdM system

the user's identity and vote to replace the linked public key currently referenced by the controller. Finally, the global registry contracts ((d) in Fig. 11.5) maps each user's uPort identifier with a hash of their identity attributes or attestations created/collected over time. uPort stores the actual identity data separately in a decentralized system called IPFS, from which the hash is generated to protect data integrity.

11.4.4.2   Sovrin

The Sovrin [32] network is another open source, public DecIdM system. It aims to provide "identity for all" by ensuring that cost does not become a hindrance for users to access its services. A simplified architecture of Sovrin is shown in Fig. 11.6.

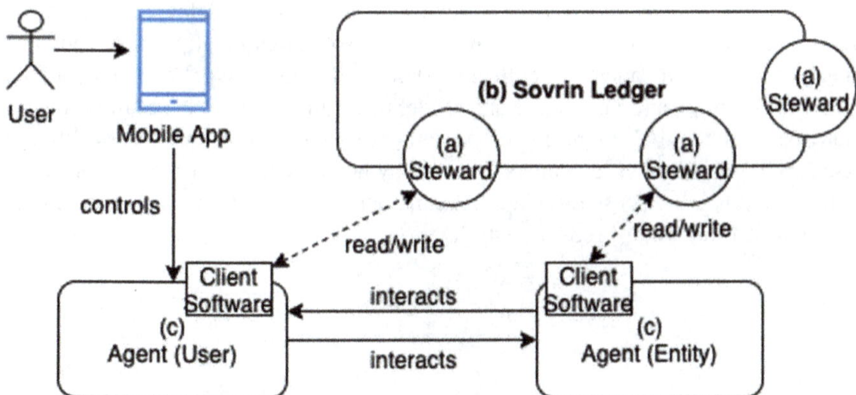

Fig. 11.6 A simplified diagram of the interactions between key components of the Sovrin Network

To accomplish this goal, Sovrin runs within a permissioned blockchain based on the Hyperledger Indy project [13]. It is governed by a group of trusted institutions called "stewards" ((a) in Fig. 11.6), who are network nodes abiding by transaction validation rules that are significantly cheaper to execute than the consensus protocols used in public blockchains such as Proof of Work [19]. As a key constituent of the Sorvin architecture, the permissioned blockchain maintains an ordered, verifiable transaction log of all accessed information and issued identity credentials and is thus known as the Sovrin "ledger" ((b) in Fig. 11.6). It is worth noting that permissions must be granted only when new stewards are introduced but are not required when users wish to query data on the ledger.

The Sovrin network grows under a "web of trust" model whereby existing stewards, such as banks, universities, or government agencies, may enroll other entities (stewards or users) to the network. Sovrin DIDs comply with the W3C DID v1.0 standard, which includes a cryptographic key pair like in uPort [7]. When a user is invited to the network by a steward, the user can then interact within Sovrin through "agents" ((c) in Fig. 11.6), which are trusted client software (e.g., a mobile app) that securely stores the user's private key(s) and communicates with other agents [21]. The user can obtain credentials from other credential issuing agents through off-chain interactions and then distribute them through their delegating agent. When key recovery is needed, Sovrin offers a few options. One approach is that the user can authorize multiple devices to access her agent, and when one of the devices is lost or stolen, that device can be revoked from the agent and thus will be denied access to her identity data [28].

11.5 The Potential of Applying Blockchain-Based DID Solutions in Health Care

As an industry with increasingly growing users, health care has been challenged with ongoing medical errors that expose patients to serious risks of delayed and incorrect treatments. A key contributor of medical errors is misidentification of patients primarily due to discrepancies in collected identifiers, with misspelled names among the most common, and ineffective use of identifiers that frequently result in double-matched patients [16]. Identities of patients are also inconsistent across different healthcare organizations as many are served by heterogeneous, centralized systems. In this section, we discuss three applications where DID solutions can mitigate the misidentification issues bred by centralization.

11.5.1   Addressing the Patient Mismatching Problem

Patient identification has been an integral part to the delivery of safe and effective healthcare services. In the US today, patient identities are managed and resolved through proprietary systems such as the Master Patient Index (MPI) and Enterprise Patient Master Index (EPMI) [15]. These systems have implemented sophisticated algorithms to locate or match an existing patient with a set of identifying information obtained from the patient during each visit. However, inconsistencies in patient identifiers, errors in data entries, and similarities between patients in a large hospital database (e.g., identical or similar names, discrepant formats of date of birth, etc) complicate the identification process and exacerbate patient mismatching, compromising the quality of care [27] and also potentially contributing to incorrect fulfillment of medications [38]. According to a previous study, ten out of 17 medical errors that contribute to approximately 195,000 deaths per year are related to incorrect identifications of patients [10].

Healthcare organizations usually assign their patients with unique patient identifiers, but those identifiers are often unrecognized by other organizations. As a patient moves from one care provider to another, their established profile from the previous provider cannot be easily transferred to the new provider, with each occurrence posing an increased chance for error in patient databases such as the MPI. This further exposes the exact problems of centralized identity management and their negative impact in the long run. On the flip side, the patient identification mismatch problem presents an opportunity to consider blockchain-based DID solutions that naturally replace a service-centric model with a user-centric model [37]. DIDs are modeled to build single, unique personal identity profiles that include a collection of verifiable and traceable issuances of credentials. This model would free patients from having to answer repeated questions during their medical visits or request access to new patient portals when transitioning between various points of care. It would also alleviate providers' workload by expediting the identity verification process and decrease patient mismatch rate due to similar or inconsistent identifiers. Moreover, it can protect patient privacy by allowing them to regain ownership of their identity.

11.5.2   Facilitating Patient Recruitment in Clinical Research Using DID

Clinical trials are an integral part of medical research that are designed to evaluate the safety and efficacy of experimental treatments and intervention methods to ultimately improve human health. In order to conduct valid, generalizable science, researchers must often acquire vast amounts of data from large patient cohorts that meet the eligibility criteria to draw statistically significant conclusions. It is expensive to correctly identify appropriate patients and then collect and manage their profiles. This expense often leads to proprietary patient profiles that are not openly shared with other interested parties, such as hospitals or research teams. In rare disease studies that already have small numbers of eligible participants who may be highly geographically dispersed, the inability to identify patients for potential recruitment is much more problematic. In reality, many trials face unfortunate delays or premature terminations due to failure to reach expected patient enrollment [5, 18, 30].

The employment of DIDs has strong potential for supporting the recruitment process that impedes the success of clinical research. First of all, although many trials are advertised publicly through various channels including study hospital's website, social media posts, or physical classifieds posts, there exists distrust from eligible participants who have not yet interacted with the hospital or the study team [23, 24]. Given the tremendous amount of public information available today, it can be hard to verify the legitimacy of its sources. However, if research entities are equipped with DIDs that are designed to be verifiable, the distrust can be easily reduced. In this scenario, the research entity would only need to publish relevant DID credentials issued by trusted authorities in order to prove the authenticity of the recruitment.

Second, the uniqueness of DIDs and their credential establishment allow for more precise and efficient identification or re-identification of participants. Because of the experimental nature of clinical trials, multiple screening phases are typically involved in each study to pre-screen volunteers for an initial assessment of their eligibility to participate, obtain their informed consent to data collection, evaluate their qualifications for the study, and then enroll them in the study. During this process, a volunteer has several contacts with the researcher, via email, by phone, or in person. It is critical that the data collected refers to and only to the same person for safety reasons and to ensure validity of the results. With a DID model, the research study would establish a credible connection record with each volunteer upon their first exchange of contact, thereby making re-identification of the same volunteer more efficient during follow-up visits (compared to assigning new artificial patient identifiers for different clinical trials). The researcher may also provide the volunteer with one or more credentials certifying partial or full participation in a trial for the volunteer to receive agreed upon reimbursement if appropriate. Additionally, the credentials would help study teams and volunteers to prevent repeated enrollment and honor any "washout" period after a trial, during which volunteers must not receive any treatments [9].

Third, DIDs can facilitate the linking of Electronic Health Records (EHRs) from multiple healthcare centers for research purposes. EHRs are widely used by providers and hospitals to document patient data and contain rich information about the diagnoses, allergies, past surgical procedures, and other medical data of patients that they service. EHRs are therefore valuable resources for identifying potential eligible participants for research trials. The trial team may receive Clinical Trial Alerts (CTAs) when there are patients with the targeted medical condition being treated at their institution [8]. However, there are two variables in the number of potentially eligible participants that can be identified: (1) the size of the patient population at that institution and (2) recency or completeness of patient records. As patients change providers, their previous medical data may or may not be transferred to the EHR system at the new provider's office, creating outdated or fragmented health records.

What this means for research is that the recruiting team may lose a portion of patients who could otherwise have been eligible participants. Currently, many EHRs (e.g., EPIC) incorporate aforementioned MPI or EMPI for patient identification, which, if substituted with DIDs, would facilitate linked data of the same patient from multiple EHR systems. Consequently, when healthcare data is more complete, the CTAs would capture more qualified patient participants to increase the statistical power of the study.

11.5.3 The Safer Use of Medical Devices

Medical devices have transformed health care as an indispensable part of treatment delivery, patient monitoring, alert automation, diagnosis, and many other medical workflows. The safety of medical devices is of paramount importance, part of which is the assurance of matching devices to the correct patients being treated. For example, high-cost drug delivery devices used during surgical procedures are shared between a number of patients. Failing to identify the correct device-patient pair can lead to fatal errors, such as incorrect or omitted drug administration, improper drug dose, etc. The errors are often due to human factors (some of which are perpetuated by the software design), such as incorrect device labeling or erroneous data entries in the device management software [2]. Most medical devices today are designed to have the ability to store control software and data. From this prospective, medical devices are natural subjects that can benefit from the use of DIDs.

A DID assigned to each device would correspond to a persistent and unique identity profile similar to any human user or entity. A software counterpart then consumes this DID so that its medical staff user is able to link its associated device with the appropriate patient being treated. The pairing is captured as an aspect or a credential of the DID profile and becomes easily verifiable. Additional information can be certified by one or more medical professionals to specify the medication type and dosage to be delivered. The incorporation of DIDs would reduce the manual effort of device labeling and provide a safer use of medical devices.

Another aspect of device safe use would be the ability to handle medical device incidents and recall in a timely manner. Although medical devices are carefully tested before being employed in care settings, errors and malfunctions can still occur especially after an extended period of usage due to unforeseen system or software failures. Some errors can be hard to detect because it is impractical for medical professionals to be constantly inspecting all devices used to monitor patients. When those errors are eventually uncovered, the devices may have been used on multiple patients. After the faulty devices are recalled, they also should be traced to identify which patients had been treated by them, and ideally, at what time window, so that any incorrectly documented data can be rectified [11]. If devices are integrated with DIDs, both information would be accessible by querying the persistent transaction logs of timestamped encounters between DID-identifiable devices and patients. The available history of device usage would facilitate more adequate and safer assessments of patients and any data affected by recalled devices.

11.5.4 Limitations Facing Current Blockchain-Based DID Solutions

Health care is an industry that hinges on serving users under rigorous regulations and protocols. Systems in use by healthcare providers today were certified to meet the same requirements prior to their adoption. As any other technology that threatens the disintermediation of legacy system components in this industry, current DID solutions face three main challenges of system scalability, service availability, and user acceptability.

First, healthcare systems must be capable of satisfying the voluminous transaction demands between providers, patients, and other key stakeholders. Virtually all those transactions involve identifying affected patients, which would be naturally recorded on the blockchain by existing DID solutions. Unlike centralized systems, blockchains operate on replicating the complete transaction history onto each maintainer node in the network. With large numbers of users whose identities are exchanged on a daily basis, the scalability of DID solutions must be assessed to ensure they do not become the bottleneck of clinical communications.

Second, although blockchain-based DID solutions remove much of the dependency from intermediaries in principle, some degree of centralization necessarily remains such as during the beginning phases of identity establishment. For instance, in order to prove an attribute of a patient's identity (e.g., insurance enrollment), the patient must create a claim upfront that is then verified by some trusted authority or notary service (e.g., the insurer). This formal verification is necessary to produce a credential ensuring non-repudiation of the patient's identity attribute. The credential will be reusable within the expiry, but its initial acquisition relies on trusted services being available. As a result, the availability of any DID system especially during its early adoption would inevitably be affected by the availability of those services.

Furthermore, the adoption of DID systems would unquestionably affect their end users. Despite the intended improvement in user experience over time, the transformation in how patient users will manage and present their identities may be disorienting at the outset. A considerable number of patients own smartphones that can connect them to DID services, but for users who are not accustomed to digital services may resist the novel concepts and shift in their identity custodian. Directly replacing existing identity solutions with the notion of decentralized identity may be limited by user acceptability. One strategy to overcome this limitation may be to concurrently operate both legacy and new solutions for a transitional period.

11.6   Conclusion

In this chapter, we explored the feasibility of employing blockchain-based identity solutions in healthcare applications. We introduced a brief history of the development of identities in the digital era that focused primarily on centralized solutions. The limitations of existing identity frameworks were clear and thus motivated the need for decentralization that would allow users to regain control of the ownership of their own identities. We then provided a systematic overview of blockchain-based identities followed by two example DID system implementations, namely uPort and Sovrin. Having uncovered the key ideas and advantages of decentralized identity management, we further identified its potential use cases in a variety of healthcare scenarios. Overall, employing DID models would alleviate much of the manual and error-prone effort involved in medical workflows and consequently reduce the opportunities for costly mistakes such as mismatching in patients or misidentification of medical devices. Nevertheless, to successfully and safely exploit the benefits of DID solutions, design decisions and assessments that focus on patient privacy and data security must be carefully considered in early stages of their adoption.

References

1. Benantar, M.: Access Control Systems: Security. Identity Management and Trust Models. Springer Science & Business Media, Berlin (2005)
2. Brixey, J., Johnson, T.R., Zhang, J.: Evaluating a medical error taxonomy. In: Proceedings of the AMIA Symposium, p. 71. American Medical Informatics Association (2002)
3. Buterin, V., et al.: Ethereum white paper: a next generation smart contract & decentralized application platform. First version **53** (2014)
4. Cameron, K., Posch, R., Rannenberg, K.: Proposal for a common identity framework: A user-centric identity metasystem http://www.identityblog.com/wp-content/images/2009/06. User-CentricIdentityMetasystem. html (2009)
5. Carlisle, B., Kimmelman, J., Ramsay, T., MacKinnon, N.: Unsuccessful trial accrual and human subjects protections: an empirical analysis of recently closed trials. Clinical Trials **12**(1), 77–83 (2015)
6. Chadwick, D.W.: Federated identity management. In: Foundations of Security Analysis and Design V, pp. 96–120. Springer, Berlin (2009)
7. Dunphy, P., Petitcolas, F.A.P.: A first look at identity management schemes on the blockchain. IEEE Secur. Privacy **16**(4), 20–29 (2018)
8. Embi, P.J., Jain, A., Clark, J., Bizjack, S., Hornung, R., Martin Harris, C.: Effect of a clinical trial alert system on physician participation in trial recruitment. Arch. Int. Med. **165**(19), 2272–2277 (2005)
9. Evans, S.R.: Clinical trial structures. J. Exp. Stroke Trans. Med. **3**(1), 8 (2010)
10. Fernández-Alemán, J.L., Señor, I.C., Lozoya, P.Á.O., Toval, A.: Security and privacy in electronic health records: A systematic literature review. J. Biomed. Inf. **46**(3), 541–562 (2013)
11. Fiedler, B.A.: Device failure tracking and response to manufacturing recalls. In: Managing Medical Devices Within a Regulatory Framework, pp. 263–275. Elsevier (2017)
12. Hardt, D. et al.: The oauth 2.0 authorization framework. Technical report, RFC 6749 (2012)
13. Hyperledger.org. Hyperledger Indy: https://www.hyperledger.org/use/hyperledger-indy
14. Jøsang, A., Pope, S.: User centric identity management. In: AusCERT Asia Pacific Information Technology Security Conference, pp. 77. Citeseer (2005)

15. Just, v, Marc, D., Munns, M., Sandefer, R.: Why patient matching is a challenge: research on master patient index (MPI) data discrepancies in key identifying fields. In: Perspectives in Health Information Management, vol. 13(Spring) (2016)
16. Lippi, G., Mattiuzzi, C., Bovo, C., Favaloro, E.J.: Managing the patient identification crisis in healthcare and laboratory medicine. Clin. Biochem. **50**(10–11), 562–567 (2017)
17. Lundkvist, C., Heck, R., Torstensson, J., Mitton, Z., Sena, M.: Uport: A platform for self-sovereign identity. https://whitepaper.uport.me/uPort_whitepaper_DRAFT20170221.pdf (2017)
18. Mahon, E., Roberts, J., Furlong, P., Uhlenbrauck, G., Bull, J.: Barriers to clinical trial recruitment and possible solutions: a stakeholder survey. Appl. Clin. Trials **24** (2015)
19. Nakamoto, S.: A peer-to-peer electronic cash system (2008)
20. Narayanan, A., Bonneau, J., Felten, E., Miller, A., Goldfeder, S.: Bitcoin and Cryptocurrency Technologies: A Comprehensive Introduction. Princeton University Press (2016)
21. Nauta, J.C., Joosten, R.: Self-sovereign identity: A comparison of Irma and Sovrin. Technical report, Technical Report TNO2019R11011 (2019)
22. Confessore, N.: Cambridge Analytica and Facebook: The Scandal and the Fallout So Far: https://www.nytimes.com/2018/04/04/us/politics/cambridge-analytica-scandal-fallout.html
23. Patel, M.X., Doku, V., Tennakoon, L.: Challenges in recruitment of research participants. Adv. Psychiatric Treatment **9**(3), 229–238 (2003)
24. Payne, J.K., Hendrix, C.C.: Clinical trial recruitment challenges with older adults with cancer. Appl. Nurs. Res. **23**(4), 233–237 (2010)
25. Rackoff, C., Simon, D.R.: Non-interactive zero-knowledge proof of knowledge and chosen ciphertext attack. In: Annual International Cryptology Conference, pp. 433–444. Springer, Berlin (1991)
26. Roman, R., Zhou, J., Lopez, J.: On the features and challenges of security and privacy in distributed internet of things. Comput. Netw. **57**(10), 2266–2279 (2013)
27. sequoiaproject.org. Cross organizational patient identity management: Challenges and opportunities: http://sequoiaproject.org/wp-content/uploads/2017/02/2017-02-22-HIMSS-2017-Patient-Matching-Challenges-and-Opportunities-v001.pdf
28. Sovrin.org. What if someone steals my phone: https://sovrin.org/wp-content/uploads/2019/03/What-if-someone-steals-my-phone-040319.pdf
29. Sporny, M., Burnett, D.C., Longley, D., Kellogg, G.: Verifiable credentials data model 1.0: Expressing verifiable information on the web. Draft **7** (2018)
30. Sullivan, J.: Subject recruitment and retention: barriers to success, Appl. Clin. Trials (2004)
31. The 1984 National Minimum Drinking Age Act. Cambridge Analytica and Facebook: The Scandal and the Fallout So Far: https://alcoholpolicy.niaaa.nih.gov/the-1984-national-minimum-drinking-age-act
32. Windley, P.J.: How Sovrin works. Windely. com (2016)
33. Wood, G., et al.: Ethereum: A secure decentralised generalised transaction ledger. Ethereum Project Yellow Paper **151**(2014), 1–32 (2014)
34. Doffman, Z.: Facebook Dark Web Deal: Hackers Just Sold 267 Million User Profiles For $540: https://www.forbes.com/sites/zakdoffman/2020/04/20/facebook-users-beware-hackers-just-sold-267-million-of-your-profiles-for-540/#7d4a52527c85
35. Zhang, P., Kamel Boulos, M.N.: Blockchain solutions for healthcare. In: Precision Medicine for Investigators, Practitioners and Providers, pp. 519–524. Elsevier (2020)
36. Zhang, P., Downs, C., Le, N.T.U., Martin, C., Shoemaker, P., Wittwer, C., Mills, L., Kelly, L., Lackey, S., Schmidt, D., et al.: Towards patient-centered stewardship of research data and research participant recruitment with blockchain technology. Front. Blockchain **3**, 32 (2020)
37. Zhang, P., Schmidt, D.C., White, J., Lenz, G.: Blockchain technology use cases in healthcare. In: Advances in Computers, vol. 111, pp. 1–41. Elsevier (2018)
38. Zhang, P., Stodghill, B., Pitt, C., Briody, C., Schmidt, D.C., White, J., Pitt, A., Aldrich, K.: Optrak: Tracking opioid prescriptions via distributed ledger technology. Int. J. Inform. Syst. Soc. Change (IJISSC) **10**(2), 45–61 (2019)

Chapter 12
Toward eHealth with Blockchain: Success Factors for Crowdfunding with ICOs

Stefan Tönnissen and Frank Teuteberg

Abstract Blockchain is seen as having high potential for the healthcare industry. In addition to the technological properties of the blockchain, crowd funding via Initial Coin Offerings (ICOs) has also become a significant way of financing for start-ups in the healthcare industry. These ICOs are significantly different from familiar funding channels and therefore require intensive consideration of the factors relevant to success. Numerous papers have explored the success factors, but without considering the specifics of the healthcare industry. We fill this research gap and, based on hypotheses and a quantitative analysis of freely available data, show the factors relevant to the healthcare industry for a successful ICO. As a result, we show clear differences to the previously known success factors for ICOs and thus prove that the success factors of ICOs require an industry-specific consideration. Start-ups receive valuable advice on how to design a successful ICO. For the scientific community, there are starting points for further research into the success of ICOs.

12.1 Introduction

The Annual European eHealth Survey [11] shows that funding opportunities are the most important challenge for healthcare providers [20]. The innovative crowdfunding method initial coin offerings (ICO) provides digital start-ups in general and in the field of health care (eHealth) in particular an opportunity to obtain financial resources via the Internet. In the fourth quarter of 2018, 209 ICOs (out of 594 ICOs) successfully raised capital [23].

The issuance of digital tokens as cryptocurrencies for value exchange [17] under these ICOs is based on blockchain technology, a technology which also promises to bring benefits as an innovative technology to many healthcare applications, such as the management of electronic medical records (EMR) of patients [3]. EMRs include private and sensitive health records of patients and are frequently shared around

S. Tönnissen (✉) · F. Teuteberg
School of Business Administration and Economics, Osnabrueck University, Katharinenstr. 1, 49069 Osnabrück, Germany
e-mail: stoennissen@uni-osnabrueck.de

© The Author(s), under exclusive license to Springer Nature Singapore Pte Ltd. 2021   209
S. Patnaik et al. (eds.), *Blockchain Technology and Innovations in Business Processes*,
Smart Innovation, Systems and Technologies 219,
https://doi.org/10.1007/978-981-33-6470-7_12

the healthcare system. Blockchain technology provides the necessary and invariable transparency for the history and use of the data [10]. So far, it is unclear what impact Initial Coin Offerings (ICOs) are having on the adaptation of blockchain in the healthcare sector and which success factors are relevant for obtaining financing capital. Therefore, in our work, we investigate the successful initial coin offerings of eHealth start-ups in the field of blockchain and the research question: *What are the success factors for crowdfunding with ICOs for eHealth start-ups?* Our work differs from previous contributions to success factors of ICOs in the form that we use the different success factors determined from empirical work on cross-industry ICOs specifically for the health sector.

To answer this research question, in the following chapter, we first discuss the theoretical foundations of eHealth, blockchain and ICOs. Based on this, we develop our research hypotheses. The following is an explanation of the research method followed by a presentation of the underlying data of our analysis. The results of the analysis of the data will be presented in the following chapter and then discussed. At the end, conclusions and implications are as follow.

12.2 Theoretical Background

12.2.1 eHealth

The term eHealth is used today in various forms [14]. Faber [14] describes eHealth as the use of emerging information and communication technologies to improve health and health care. In its resolution WHA58.28 (2005), the World Health Organization (WHO) use the term eHealth: "eHealth is the cost-effective and secure use of information and communications technologies in support of health and health-related fields, including healthcare services, health surveillance, health literacy, health education, knowledge and research" [34]. For Liu et al. [26], eHealth is a technology that is becoming more and more important, which, in addition to electronic health records (EHR), also includes electronic medical records (EMR) and the exchange of real-time health data from various body sensors. Eysenbach [13] complements existing definitions by describing it as the global interconnection of medical informatics, public health and economy through the Internet with the aim of improving health care locally, regionally and globally through the use of information and communication technologies. We follow Eysenbach's [13] approach in our work, because the global and networked view of health care helped by information and communication technologies and describes most closely the phenomenon of blockchain as the technological basis of an initial coin offer.

12.2.2 Initial Coin Offerings (ICOs) and Blockchain

An initial coin offering (ICO) is a new form of crowdfunding based on blockchain technology. ICOs have existed since about 2013 and only with the popularity of the Ethereum blockchain in 2017 did they spread worldwide [4]. The term ICO is based on the term initial public offering (IPO) [1], with the difference that, in contrast to an IPO, no share of the company is sold, but a token, in exchange for a future sale performance or a future product [7]. The popularity of ICOs for both investors and digital start-ups is highly correlated with blockchain technology. The blockchain is a distributed ledger technology and since the introduction of bitcoins 10 years ago the technological basis of numerous digital start-ups. The blockchain is a peer-to-peer network in which the data is distributed decentrally across all participants in the network. This eliminates the need for an intermediary, unlike today where an intermediary plays a central role in many business models between service providers and service users. The data is encrypted by concatenating it in chronological order based on transactions between peer-to-peer network participants and stored in vari-ablyon all participants' computers [30]. In order to collect data from a transaction between two participants, a consensus mechanism must be used to confirm the accuracy of the transactions and establish an agreement between all participants in the blockchain network on the correct status of the data. This ensures that the data is the same at almost the same time on all nodes in the peer-to-peer network [33]. A block of data on the blockchain contains a timestamp and the hash value of the previous data blockchain addition to the data itself. As a result, the data blocks are chained together and protected against subsequent changes [21].

The popularity of ICOs for both investors and technology start-ups is highly correlated with the evolution of blockchain technology. Initially, the entire ICO process on the decentralized blockchain is completely anonymous worldwide. Because of the lack of intermediaries in a blockchain, ICO-funded companies do not need to partner with investment banks, financial services providers, or crowdfunding platforms. Thus, they not only dispense with the extensive rules of these intermediaries, but also the costs they incur [7]. Furthermore, ICOs are not yet subject to uniform regulations and can therefore provide fast liquidity for an ICO by real-time processing on the blockchain [4]. The investor benefits from the ICO by not only financing an innovative project and benefiting from future benefits through a token exchange, but also from the speculative potential to increase the value of the tokens [7] coupled with the rapid generation of liquidity [4].

The ICO market is unregulated and so far there is no central source for ICO data. Accordingly, there is no platform for compulsory registration of ICOs [17]. The data provided on the Internet is based on public sources and is not subject to a reporting standard [12]. Well-known Web sites such as Coinschedule or ICObench are tracking the emerging ICOs, but they often rely on user input 17. Mastercoin's first initial coin offerings were in 2013, so the ICO market is still quite a young market [31].

12.2.3 Signaling Theory

A rational investment decision is made on the basis of reliable information [28]. For an ICO, however, the potential investor initially relies merely on information from the entrepreneur, as many ICOs are carried out at a very early stage in order to finance the development of an application or service. This information asymmetry must be overcome by the entrepreneur in order to be able to achieve the greatest possible financial success. Spence's signaling theory [32] states that an entrepreneur sending signals to a potential investor can result in greater financing success [2]. However, if the potential investor can draw on information from third parties in addition to these signals from the entrepreneur for their investment decision, this is known as the "certification hypothesis" [6] which is an extension of the signaling theory.

12.3 Research Method

To establish our hypotheses, we first analyze related work on the success criteria for ICOs and success-determining variables. We then perform a quantitative analysis based on freely available data from past ICOs to determine the success factors for an ICO in the eHealth area. These success factors are assigned to either the signaling theory of Spence [32] or the certification hypothesis of Booth and Smith [6].

12.3.1 Related Work

We conducted a literature search on Google Scholar in November 2018 and found six related contributions. We first examined this related work to determine the criteria the authors set for a successful ICO. The result shows very different criteria, for example, a successful ICO is defined as the fulfillment of minimum funding requirements [1]. Other authors define success as the active trading of tokens on the stock exchanges [4], various liquidity ratios after six months trading on the stock exchanges [22] or an absolute financing amount of $200,000 as a threshold [16]. One author dispenses with criteria for success and includes all ICOs in his analysis 17.

Our contribution aims to gain insight into the adaptation of blockchain in eHealth. For this reason, we reject the criteria of success in the contributions of Amsden and Schweizer [4] and Feng et al. [15] and classify as successful all ICOs that were able to collect via the ICO capital, thus, exceeding the soft cap as a minimum funding target. A soft cap is when the start-ups set a minimum funding amount that must be achieved to finance the project. If the minimum amount of financing is not reached, investors get their money back [4].

The related contributions used between 7 and 36 variables for the determination of success factors, with a mean of 20.7 and a standard deviation of 9.2. For our analysis, we have adopted the following common success factors based on related work as shown in Table 12.1.

Of the common success factors as shown in Table 12.1, 37.5% (three factors) can be assigned to the certification hypothesis and 62.5% (5 factors) to the signaling theory. If the signalers have a privileged view of the success factor, then it is an association to the signaling theory. If it is not, then it connects to the certification hypothesis.

The work done so far on the periods 2017 and 2018 shows a focus on the information provided by the entrepreneur, such as the whitepaper explaining the business model or the status of the project documented on GitHub.

12.4 Research Hypotheses

An understanding of the determining factors for a successful ICO in the healthcare industry will help future entrepreneurs to generate the criteria for their project that are relevant for successful financing. An ICO creates large information asymmetries between the entrepreneur and the investor [1], although today the Internet provides a range of different information from different sources about an ICO. Some information, such as whitepapers describing the business model, source code published in GitHub or the use of social media channels such as Twitter, are provided by the entrepreneur to the potential investor. In addition, there is relevant information from third parties which can be incorporated into a potential investor's investment decision, such as a rating onICObench.com or the number of followers on Telegram or Twitter. The uncertainties for the investor in an ICO arise mainly for three reasons. First, the blockchain technology is complicated and the business models of the ICO start-ups are incomprehensible to many people. Second, the ICO start-ups are at an early stage of development and often have no product or service yet. Third, the available information is low due to a lack of regulation or disclosure requirements. The signaling theory for the reinforcement of information as signals by the investor as well as the certification hypothesis for the reinforcement of information by third parties provide the framework for explaining which factors are relevant for start-ups for the acquisition of funds [17]. Using these different information sources based on our findings in Table 12.1, our hypotheses are first structured on information from the entrepreneur to follow the signaling theory, and second on information from third parties to follow the certification hypothesis.

H1. Information from the entrepreneur about a future ICO influences the probability of success of the ICO.

We further specify this hypothesis. The whitepaper is typically an explanation of the business idea and sometimes a technical description [9]. The importance of a whitepaper for the success of an ICO is shown by Cerchiello and Toma [9], Fisch [17], and Howell et al. [22]. Thus, we formulate the following hypothesis: *H1a. The presence of a whitepaper positively influences the probability of success.*

Table 12.1 Success factors from related work with assignment to signaling theory (ST) or certification hypothesis (CH)

Success factors	CH	ST	Sources/results
Expert ratings	X		Fenu and Tonelli [16]/The relative p-value is .0001 and the coefficient value and the Z-Wald value are 1.6991 and 3.84 Feng et al. [15]/ICO projects with a higher rating raised more funding
Presence of a whitepaper		X	Cerchiello and Toma [9]/The positive coefficients 3.113 showing their impact on increasing the probability of success of an ICO Fisch, [17]/The coefficient of a whitepaper is significant ($p < .01$) and positive Howell et al. [22]/A whitepaper has large, positive coefficients
Number of team members		X	Cerchiello and Toma [9]/Number of elements of the team is highly significant and positive
Publication of the code on GitHub		X	Adhami et al. [1]/The coefficient is positive and highly significant (p-value <1%) Amsden and Schweizer [4]/The variable GitHub is positively correlated with having tradable tokens Fisch [17]/GitHub is insignificant, code on GitHub does not significantly influence the amount of funding raised Howell et al. [22]/No significant relationship between commits (the number of revisions to the repository) and liquidity
Follower Telegram	X		Cerchiello and Toma [9]/The positive coefficients 1.917 show their impact on increasing the probability of success of an ICO Amsden and Schweizer [4]/The variable Telegram is positively correlated with having tradable tokens Howell et al. [22]/Having a Telegram messaging group has large, positive coefficients
Follower Twitter	X		Howell et al. [22]/A Twitter account has no predictive power
Type of platform		X	Amsden and Schweizer [4]/Negative correlation with the Ethereum platform Fenu and Tonelli [16]
Presence of a minimum viable product		X	Adhami et al. [1]/The coefficient is positive and highly significant Howell et al. [22]/Having a utility value has positive coefficients

Cerchiello and Toma [9] emphasizes that the number of employees is a strong signal to the investor and follows the view of the resource base. Our hypothesis is therefore: *H1b. The number of team members has a positive influence on the probability of success.*

If code and updates are published on GitHub, potential investors can review the code and understand the progress of the token offer [4]. The importance of these variables has been demonstrated in the work of Adhami et al. [1], Amsden and Schweizer [4], Fisch [17], and Howell et al. [22]. *H1c. The availability of the source code on GitHub has a positive influence on the probability of success.*

A minimum viable product (MVP) is a working prototype. Adhami et al. [1] and Howell et al. [22] identified the presence of an MVP as a critical success factor for an ICO in many cases. This leads to our next hypothesis: *H1d. The presence of an MVP positively influences the probability of success.*

The buyer of an ICO token usually acquires the right to access platform services in 68.0% of the cases [1]. Therefore, the type of the platform is a relevant information for the investor. Fenu and Tonelli [16] and Amsden and Schweizer [4] show that the platform used for the ICO is an essential feature of success in their analysis. The hypothesis derived from this is: *H1e. The selected blockchain platform has a positive influence on the probability of success.*

In addition to hypothesis H1 about information from an entrepreneur, information from third parties is also relevant for rational decision making about an investment. We therefore define our second hypothesis as follows:

H2. Information from third parties about a future ICO influences the probability of success of the ICO.

We also further specify this hypothesis as follows. Communication via social media channels enables potential investors to communicate directly with an ICO start-up [4]. Cerchiello and Toma [9] as well as Amsden and Schweizer [4] and Howell et al. [22] demonstrate the great importance of these communication channels. This leads to our next two hypotheses: *H2a. The number of followers in Telegram has a positive influence on the probability of success* and *H2b. The number of followers on Twitter has a positive influence on the probability of success.*

Fenu and Tonelli [16] demonstrate the importance of ratings from third parties for the success of an ICO the following hypothesis is therefore: *H2c. The rating of the ICO positively influences the probability of success.*

From our hypotheses and their specifications, we develop the following conceptual model for successful ICOs in the healthcare industry (Fig. 12.1).

12.4.1 Data and Variables

On March 2, 2019, we searched the Internet database www.icobench.com for the status "ended" in the "health" category under initial coin offerings (ICOs), scoring 149 hits. This popular Internet database [9] as well as Fenu and Tonelli [16] provides

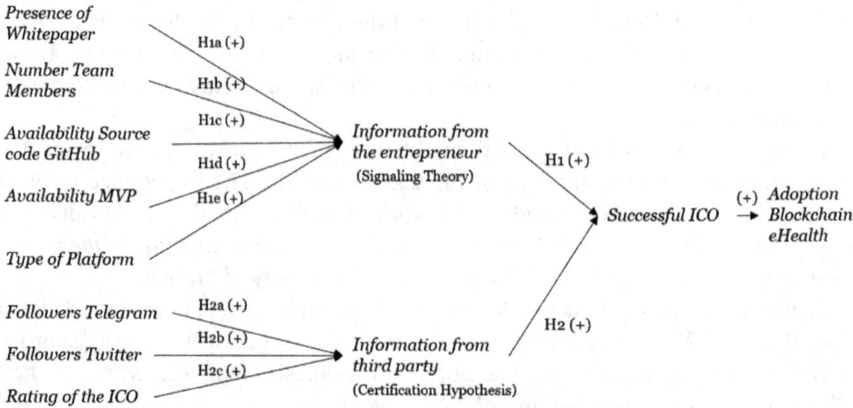

Fig. 12.1 Conceptual model for successful ICOs in eHealth

the most extensive information on ICOs [4] that have happened since September 2015. Other possible statuses for ICOs are "upcoming" and "ongoing". We focus on completed ICOs in order to answer our research questions and test for successful ICOs. Of the 149 ICOs selected, 15 were completed in 2017, 124 in 2018 and 10 in 2019. According to www.icobench.com, there were no ICOs in the health category before 2017. The data relevant for our analysis has been taken from the Internet database www.icobench.com.

12.4.1.1 Dependent Variable: eHealth Success

The ICO active dependent variable has a value of 1 to indicate that a successful ICO took place, and a value of 0 if the ICO was unsuccessful. The success of an ICO is initially measured by the value of tokens sold, also known as "raised" or "value of tokens sold in ICO." We were able to determine the relevant data from www.ico bench.com. If the data was unclear, we took data about the price changes in the tokens as well as circulating supply from www.coinschedule.com. Price trends on the crypto exchanges indicate active tokens with a corresponding round of purchases and sales. On average, 36.2% of the ICOs are successful and have been able to generate funds for the growth of start-ups through the ICO.

12.4.1.2 Independent Variables

The independent variables are the success factors derived from the related work and are presented in Table 12.1, which we explain below.

We have included the ratings of the ICOs from the ICObench database and have combined an evaluation algorithm with more than 20 different criteria and the ratings of independent experts [24] in the variable rating. The variable has a value between zero and five. Fenu and Tonelli [16] demonstrate the importance of ratings for the success of an ICO.

If there is a whitepaper, we set the whitepaper variable to 1 and the value to 0 if there is no whitepaper. A whitepaper outlines the business model [9] to the potential investor from an economic and technical point of view and explains how to use tokens including benefits for the owners [22]. The importance of such a whitepaper for the success of an ICO is shown by Cerchiello and Toma [9], Fisch [17] and Howell et al. [22].

The number of team members is set in the TeamSize variable from the information on www.icobench.com. Cerchiello and Toma [9] demonstrate that the number of team members is highly significant for the success of an ICO.

If the project and the source code is published on GitHub, the variable GitHub is set to 1; if there is no data, the variable is set to 0. GitHub is a source code-publishing Web site and supports collaboration between software developers [1, 29]. The importance of these variables has been demonstrated in the work of Adhami et al. [1], Amsden and Schweizer [4], Fisch [17], and Howell et al. [22].

Communication via social media channels enable potential investors to communicate directly with an ICO start-up [4]. The Telegram variable records the number of followers on social media channel Telegram and the Twitter variable for social media channel Twitter. Cerchiello et al. [9] as well as Amsden and Schweizer [4] and Howell et al. [22] demonstrate the great importance of these communication channels.

A minimum viable product (MVP) is a working prototype. Adhami et al. [1] and Howell et al. [22] identified the presence of an MVP as a critical success factor for an ICO in many cases, a minimum viable product (MVP) is not available at the time of the ICO [11], since it is precisely the financial resources from the ICO that are needed for this very development. In this case, the purchase of a token is a decision to potentially use a digital start-up idea rather than a decision to use a specifically available service.

The blockchain platform used for the ICO is stored in the platform variable. Fenu and Tonelli [16] and Amsden and Schweizer [4] show that the platform used for the ICO is an essential feature of success in their analysis.

In addition to the variables relevant to the hypotheses, we also include three additional control variables to verify our investigations. First, the bonus as an instrument to attract future investors and to bring in investment (bonus variable) [1]. Many start-ups award bonuses in the form of free tokens in advance of an ICO for services or benefits. Basically, these bonuses can be differentiated into volume bonuses, as a volume discount for the purchase of tokens from a certain amount, a bonus for the early purchase of tokens in the context of a pre-ICO and a bonus as a reward for attracting additional investors. In addition to the early financing success for the start-up, the investor benefits from a lower investment risk [4]. In addition to the incentives provided by various bonus programs, digital start-ups often offer bounty programs (bounty variable) in which future investors and users of the products or services receive a reward in exchange for taking on specific tasks. The incentives may consist of cash rewards and free or discounted tokens that can be redeemed later when the tokens are listed on a stock exchange [18]. The Internet database ICObench specifies "bonus available" as well as "bounty available" for each ICO. If the characteristic applies to the variable, we set the variable to 1, and if not, we set it to 0. Up-to-date information is required on the homepage of the ICO to overcome the information asymmetries between entrepreneur and investor. The Alexa variable measures the number of average visitors per day (daily page views per visitor), which we determine from https://www.alexa.com/ [15].

12.5 Results and Discussion

Our evaluation of 149 ICOs has yielded 54 successful ICOs (36.2%). Based on our data, we conduct a study to determine which factors influence the success of an ICO [16].

12.5.1 Descriptive Statistics of the ICO Data

The 149 ICOs included in our analysis are distributed between 9 countries; the USA (29x, 19.5%), the UK (15x, 10.1%), Estonia (11x, 7.4%), Switzerland (11x, 7.4%), Singapore (9x, 6.0%), Russia (8x, 5.4%), the Cayman Islands (5x, 3.4%), the Czech Republic (5x, 3.4%), and Hong Kong (5x, 3.4%).

At 87.2%, the blockchain Ethereum clearly dominates the platforms. This follows the trend on ICObench, that out of 5368 ICOs, 4722 were realized on the Ethereum blockchain, accounting for 87.9%. The success of Ethereum is attributed to the ERC20 standard used for the tokens [5] and that ICOs are simple to set up and implement [8, 9].

The average rating of health ICOs is 3.142, below the average rating of 3.6 out of 209 successful ICOs across all sectors in Q4 in 2018 (according to ICObench.com).

Of the 149 ICOs, 32 ICOs (21.5%) have a minimum viable product (MVP). This is low compared to current ICOs across all industries, accounting for 52% (according to ICObench.com).

Tables 12.2 and 12.3 show the monovariate statistics and the correlation matrix for the variables in our analysis as well as the variance inflation factor (VIF). The VIF value indicates the degree of collinearity or multicollinearity among the independent variables. A VIF score above 4.0 indicates high multicollinearity among covariates [19], which we cannot determine from the VIF scores.

The correlations between variables (see Table 12.3) show no high values between the variables. The highest values are between the rating variable and the Team-Size variable, at.604, and the KYC variable and the rating tag, at .634. All other correlations between the variables are lower than .450.

Table 12.2 Descriptive statistic and VIF

Variable	N	Minimum	Maximum	Average	Standard deviation	VIF
ICO-active	149	0	1	.75	.433	
Bonus	149	0	1	.52	.501	1.252
Bounty	149	0	1	.34	.476	1.242
MVP	149	0	1	.21	.412	1.548
KYC	149	0	1	.32	.466	1.918
Rating	149	1.3	4.5	3.142	.6742	2.711
TeamSize	149	0	40	13.60	7.789	1.627
Telegram	104	0	73,628	8352.64	12786.279	1.567
Twitter	106	0	37,607	4145.48	6599.657	1.319
Alexa	149	.00	6.40	1.1430	1.06405	1.193
whitepaper	149	0	1	.95	.212	1.389
GitHub	149	0	1	.53	.501	1.214

Table 12.3 Result of the correlation analysis

	1	2	3	4	5	6	7	8	9	10	11	12
ICO-active 1	1											
Bonus 2	.114	1										
Bounty 3	−.044	.188	1									
MVP 4	.150	.211	.243	1								
KYC 5	.299	.310	.302	.348	1							
Rating 6	.374	.257	.247	.434	.634	1						
TeamSize 7	.195	.150	.186	.406	.405	.604	1					
Telegram 8	.138	.189	.102	.329	.345	.341	.284	1				
Twitter 9	.251	.102	.008	.027	.120	.279	.108	.381	1			
Alexa 10	.204	.019	.065	.250	.211	.283	.320	.263	.204	1		
whitepaper 11	−.031	.039	.093	−.038	.082	.326	.083	.074	.112	−.084	1	
GitHub 12	.066	.166	.141	.132	.234	.359	.193	−.072	.120	.056	.109	1

12.5.2 Multivariate Analysis of the Factors Influencing the Success

We perform a logistic regression analysis with the dependent and binary variable ICO active using SPSS version 24. The general form of logistic regression as a relationship between a single binary variable and a set of independent variables can be described as follows [19] (Table 12.4):

Table 12.4 Result of logistic regression

Attribute	Variable	Regression coefficient B	Wald	Sig.	Exp(B)	ST	CH
H1a	whitepaper	−.237	.019	.891	.789	X	
H1b	TeamSize	.003	.006	.937	1.003	X	
H1c	GitHub	−.589	1.178	.278	.555	X	
H1d	MVP	−.396	.391	.532	.673	X	
H1e	Platform	−.449	.292	.589	.638	X	
H2a	Telegram	.000	1.093	.296	1.000		X
H2b	Twitter	.000	2.444	.118	1.000		X
H2c	Rating	1.849	7.471	.006	6.353		X
Control	Bonus	.356	.438	.508	1.428	X	
Control	Bounty	−.806	2.095	.148	.446	X	
Control	Alexa	.238	1.148	.284	1.268		X
Constant	Constant	−5.909	7.031	.008	.003		

(ST = signaling theory/CH = certification hypothesis)

Table 12.5 Model summary

Step	−2 Log-likelihood	Cox and Snell R-quadrat	Nagelkerkes R^2
1	112.523	.252	.338

$$Y_1 = X_1 + X_2 + X_3 + \cdots + X_n$$
(binary nonmetric)   (nonmetric and metric)

The significance of the variables on the independent variable is then verified using the Wald test. If the Wald reaches a significance level of at least .05, this means a z-value of greater than 1.96 or less than −1.96 (for a standard normally distributed test statistic) [19, 25]. The result of our logistic regression shows the Bounty (2.095), Rating (7.471) and Twitter (2.444) variables with their corresponding significance values (Table 12.5).

The model summary shows the quality of our model with an acceptable Nagelkerkes R^2 of .338.

12.6 Discussion

The logistic regression analysis (see Table 12.6) shows that one hypothesis is supported, three hypotheses have a low significance and four have no significance. Only the rating variable shows sufficient significance to support our H2C hypothesis. The Telegram and Twitter variables show results for hypotheses H2a and H2b but with a weak significance. Hypotheses H2a, H2b and H2c are confirmed by the logistic

Table 12.6 Hypotheses with relative p-value and result

Hypothesis	Relative p-value	Result
H1. Information from the entrepreneur about a future ICO influences the probability of success of the IC	–	Unsupported
H1a. The presence of a whitepaper positively influences the probability of success	.891	Unsupported
H1b. The number of team members has a positive influence on the probability of success	.937	Unsupported
H1c. The availability of the source code on GitHub has a positive influence on the probability of success	.278	Weakly supported
H1d. The presence of an MVP positively influences the probability of success	.532	Unsupported
H1e. The selected blockchain platform has a positive influence on the probability of success	.589	Unsupported
H2. Information from third parties about a future ICO	–	Supported
H2a. The number of followers in Telegram has a positive influence on the probability of success	.296	Weakly supported
H2b. The number of followers on Twitter has a positive influence on the probability of success	.118	Weakly supported
H2c. The rating of the ICO positively influences the probability of success	.006	Supported

regression and thus the superior hypothesis H2 is confirmed. Information from third parties about a future ICO influences the probability of success of an ICO. On the other hand, the superordinate hypothesis H1, i.e., information from the entrepreneur on a future ICO does not influence the probability of success of the ICO since four out of five hypotheses are not supported by the result, and hypothesis H1c has only a weak significance.

Fenu and Tonelli [16] obtained a highly significant value for the rating variable with a relative p-value of .0001 and coefficient values and Z-Wald values of 1.6991 and 3.84, respectively. They point out that ICObench.com's rating is a reliable indicator of the potential success of an ICO. Feng et al. [15] also show the importance of ratings for the amount of funding of an ICO. The data collected in the histograms (see Fig. 12.2) show an average rating of 3.48 for the successful ICOs (right panel) and a significantly lower 2.95 for the unsuccessful ICOs (left panel).

Surprisingly, other contributions do not confirm the importance of the rating for the success of an ICO, or they ignore the influence of the rating. Adhami et al. [1], for example, ignore external information and only examine variables that focus on signaling theory.

In three of our contributions (see Table 12.1), the presence of a whitepaper is a significant factor for a successful ICO. We cannot confirm this result on the basis of our evaluation, because the relative p-value of .891 does not support our hypothesis H1a. Amsden and Schweizer [4] point out that the lack of a consistent format for whitepapers makes it difficult to compare ICO projects. The existence of a whitepaper

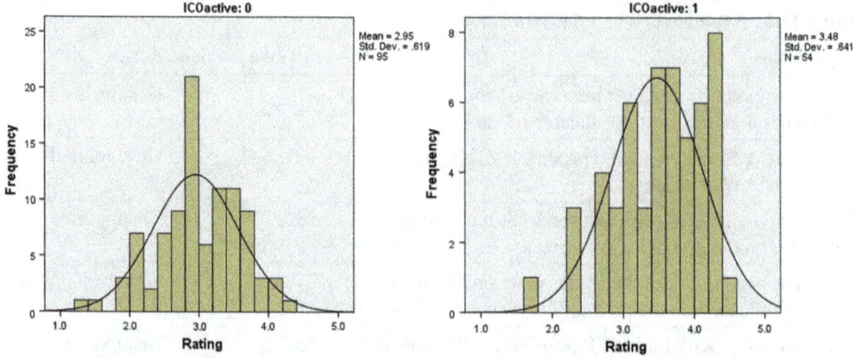

Fig. 12.2 Histograms for the ratings in the case of failure (left) and success (right) of an ICO

is not enough to eliminate investor uncertainty. The smaller the number of pages in the whitepaper, the greater the uncertainty [4]. Out of our 149 ICOs, 142 include a whitepaper, regardless of scale and quality (95.3%).

The number of team members (TeamSize variable) has no significant impact on the success of an ICO and does not support our hypothesis H1b. The successful ICOs have an average of 15.61 team members (right in Fig. 12.3), while the unsuccessful ICOs have an average of 12.46 team members (left in Fig. 12.3). Cerchiello and Toma [9] come to a different conclusion, making the success of an ICO highly dependent on the number of the team members. In their work, they refer to the analysis of 120 ICOs, but their success criterion is determined by a text analysis with a focus on fraud. The other authors do not mention the importance of the number of team members. This is all the more astonishing as the success of a young company is often highly dependent on the resources available, and these are resource-based resources as well as financial resources [27].

The availability of source code on GitHub is weakly significant (p-value 1.178) for the success of an ICO. Adhami et al. [1], however, show a high degree of significance

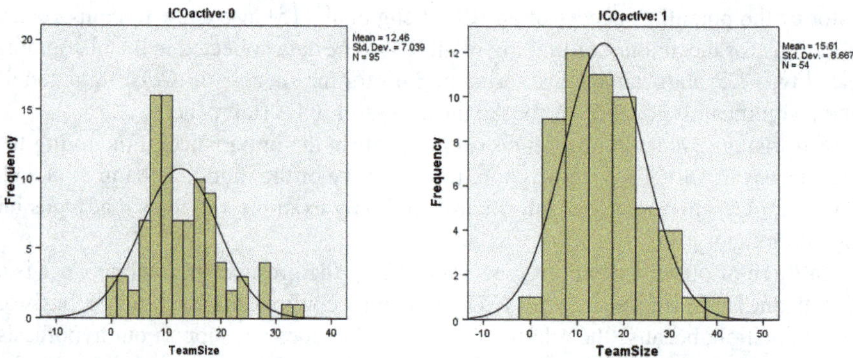

Fig. 12.3 Histograms for TeamSize in the failure (left) and success (right) of an ICO

(p-value 1.4187) for the availability of source code on GitHub and emphasize its importance for technology-savvy investors. For Amsden and Schweizer [4], the lack of source code on GitHub leads to greater investor uncertainty, which in turn has a negative effect on a successful ICO. However, Fisch [17] concludes that the disclosure of source code on GitHub has no impact on the success of an ICO. He justifies this by pointing out that the disclosure of source code does not entail any significant expense for the entrepreneur, and therefore the investor cannot assess the quality of the source code [17].

The presence of an MVP shows no significant effect on the success of an ICO (p-value .391). Adhami et al. [1] investigate the influence of a concrete token service and establish a high degree of significance. They believe that an MVP gives an investor the opportunity to use a particular service and increases the likelihood of an investment [1].

The type of blockchain platform has, according to our evaluation, no significant impact on the probability of success of an ICO. Amsden and Schweizer [4] even show a negative correlation for the Ethereum platform. This result, which was based on some of the largest ICOs, they explain as being due to Ethereum's limitations. Fenu and Tonelli [16], however, show that the Ethereum platform has high significance for the success of an ICO.

The numbers of followers in both Telegram (Wald 1093 and Sig., .296) and Twitter (Wald 2.444 and Sig., .118) are significant for the success of an ICO, even if the scores are not particularly high. Amsden and Schweizer [4] emphasize the need to communicate via Telegram for the success of an ICO. Both Cerchiello and Toma [9] and Howell et al. [22] confirm the high significance of both communications via Telegram and via Twitter. Fisch [17] explains the high significance by the fact that activities on Telegram and Twitter and regular updates are important in order to reduce existing information asymmetries and to attract potential investors [17].

Our three control variables show a differentiated picture. The bonus variable is not significant (Wald, 438 and Sig., 508) for the success of an ICO, as confirmed by Adhami et al. [1]. Amsden and Schweizer [4], however, argue that bonuses are very important and the bonus associated with a pre-ICO is a strong signal for future investors. The Alexa (Wald 1.148 and Sig., .284) and bounty (Wald 2.095 and Sig., .148) variables are only weakly significant.

12.7 Conclusion

In this paper, we examine the success factors for initial coin offerings (ICOs) in the health sector. ICOs have been a popular tool for crowdfunding digital start-ups over the Internet for about three years. The Ethereum blockchain platform allows fast and easy generation of tokens, which are exchanged over the Internet by investors against cryptocurrencies. The boom phase for ICOs in 2017 has been the subject of numerous scientific studies aimed at identifying the success factors for successful

ICOs. Not every ICO was able to achieve its financing goals. An ICO needs to overcome information asymmetries between the entrepreneur and the investor in order to generate an incentive to invest. To make a rational investment decision, the investor has information from the entrepreneur such as a whitepaper (signaling theory) and information from third parties such as ratings (certification hypothesis). The risks for the investor are high because, in contrast to an initial public offering (IPO), ICOs are not subject to regulation and supervision—a trusted middleman, such as a bank, is missing. We have taken the success factors and incorporated them into two major and eight sub-hypotheses for determining the success of health-related ICOs. Our collection and analysis of data from 149 ICOs in the health sector shows that the rating of an ICO has a significant impact on the success of an ICO, and the release of source code on GitHub and the number of followers on Telegram and Twitter have a weakly significant influence. The main hypothesis that information from third parties positively influences the probability of a successful ICO is supported by our data, while the information provided by the entrepreneur on a future ICO does not increase the probability of success and thus does not support our second main hypothesis. This result is surprising, as previous work highlights the relevance of this information. This could be explained by the healthcare industry having different success factors than the ones already determined or the price decline for bitcoins and other cryptocurrencies could be ushering in a transition from a bull market to a bear market with changed success factors. Due to numerous fraudulent ICOs in the recent past, the information given by (potentially dishonest) entrepreneurs (whitepaper, homepage, etc.) should be classified as weak signals to the investor and higher weight should be given to information from third parties. With every fraud case, the call for regulation is getting louder. Research is required into the design of legal and tax regulations, taking into account a global distributive ledger.

References

1. Adhami, S., Giudici, G., Martinazzi, S.: Why do businesses go crypto? An empirical analysis of initial coin offerings. J. Econ. Bus. (2018). https://doi.org/10.1016/j.jeconbus.2018.04.001
2. Ahlers, G.K.C., Cumming, D.J., Guenther, C., Schweizer, D.: Signaling in equity crowdfunding. Entrepreneurship Theory Pract. **39**, 955–980 (2015)
3. Alonso, S.G., Arambarri, J., López-Coronado, M., de la Torre Díez, I.: Proposing new blockchain challenges in eHealth. J. Med. Syst. **43**, 64 (2019). https://doi.org/10.1007/s10 916-019-1195-7
4. Amsden, R., Schweizer, D.: Are blockchain crowdsales the new "gold rush"? Success Determinants of Initial Coin Offerings (2018). https://ssrn.com/abstract=3163849. Retrieved 29 Nov 2018
5. Benedetti, H.E., Kostovetsky, L.: Digital Tulips? Returns to Investors in Initial Coin Offerings (2018). http://dx.doi.org/10.2139/ssrn.3182169
6. Booth, J.R., Smith, R.L.: Capital raising, underwriting and the certification hypothesis. J. Financ. Econ. **15**(1986), 261–281 (1986)
7. Brenneke, M., Fridgen, G., Guggenberger, T., Radszuwill, S.: Blockchain and initial coin offerings: blockchain's implications for crowdfunding. In: Treiblmaier, H., Beck, R. (eds.)

Business Transformation through Blockchain. Springer International Publishing (2018). https://doi.org/10.1007/978-3-319-98911-2

8. Catalini, C., Gans, J.S. Initial coin offerings and the value of crypto tokens. MIT Sloan School Working Paper 5347-18 (2018). https://ssrn.com/abstract=3137213. Retrieved 13 Feb 2019

9. Cerchiello, P., Toma, A.M.: ICOs success drivers: a textual and statistical analysis. DEM Working Paper Series (2018). ISSN: 2281-1346, http://economiaweb.unipv.it/wp-content/uploads/2018/01/DEMWP0164.pdf. Retrieved 03 Jan 2019

10. Dubovitskaya, A., Xu, Z., Ryu, S., Schumacher, M., Wang, F.: In: E. Begoli et al. (eds.) How Blockchain Could Empower eHealth: An Application for Radiation Oncology. DMAH 2017, LNCS 10494, pp. 3–6. Springer International Publishing AG (2017). https://doi.org/10.1007/978-3-319-67186-41

11. EY.: EY study: Initial Coin Offerings (ICOs). The class of 2017—one year later (2018) https://www.ey.com/Publication/vwLUAssets/ey-study-ico-infographics/$FILE/ey-study-ico-infographics.pdf. Retrieved 29 Nov 2018

12. EYGM Limited.: EY research: initial coin offerings (ICOs) (2018) https://www.ey.com/Publication/vwLUAssets/ey-research-initial-coin-offerings-icos/$File/ey-research-initial-coin-offerings-icos.pdf. Retrieved 11 Dec 2018

13. Eysenbach, G.: What is e-health? J. Med. Int. Res. $\mathbf{3}$(2), e20 (2001). https://doi.org/10.2196/jmir.3.2.e20

14. Faber, S.: Factors influencing eHealth adoption by Dutch hospitals: An empirical study (2014). https://repository.tudelft.nl/islandora/object/uuid%3A2ad1d153-2bf4-4de7-831d-b9b4b159ee25. Retrieved 13 Feb 2019

15. Feng, C., Li, N., Lu, B., Wong, M.H., Zhang, M.: Initial Coin Offerings, Blockchain Technology, and Voluntary Disclosures (2018) https://papers.ssrn.com/sol3/papers.cfm?abstract_id=3256289. Retrieved 26 Mar 2019

16. Fenu, G., Tonelli, R.: The ICO phenomenon and its relationships with ethereum smart contract environment. In: Tonelli, R., Ducasse, S., Fenu, G., Bracciali, A. (eds.) IEEE 1st International Workshop on Blockchain Oriented Software Engineering (IWBOSE), Proceedings (2018)

17. Fisch, C.: Initial coin offerings (ICOs) to finance new ventures. J. Bus. Ventur. $\mathbf{34}$(2019), 1–22 (2019). https://doi.org/10.1016/j.jbusvent.2018.09.007

18. Frankenfield, J.: Bounty Programs (ICO) (2018) https://www.investopedia.com/terms/b/bounty-programs-ico.asp. Retrieved 12 Feb 2019

19. Hair, J.F., Black, W.C., Babin, B.J., Anderson, R.E. Multivariate Data Analysis. Pearson Education Limited (2014)

20. HIMSS.: Annual European eHealth Survey 2018. eHealth TRENDBAROMETER Q4/2018 (2018). https://www.himss.eu/content/annual-european-ehealth-survey-2018. Retrieved 11 Feb 2019

21. Holotiuk, F., Pisani, F., Moormann, J.: The impact of blockchain technology on business models in the payments industry. In: Leimeister, J.M., Brenner, W. (eds.) Proceedings der 13. International en Tagung Wirtschaftsinformatik (WI 2017). St. Gallen (2017)

22. Howell, S.T., Niessner, M., Yermack, D.: Initial coin offerings: financing growth with cryptocurrency token sales. Finance Working Paper N° 564/2018 (2018) https://ecgi.global/sites/default/files/working_papers/documents/finalhowellniessneryermack.pdf. Retrieved 27 Nov 2018

23. ICObench.: Stats and Facts. ICO Market Reports (2019a). https://icobench.com/stats. Retrieved 11 Feb 2019

24. ICObench.: Rating Methodology (2019b). https://icobench.com/ratings. Zugriff am 13.02.2019

25. Kutner, M.H., Nachtsheim, C.J., Neter, J., Li, W.: Applied Linear Statistical Models, 5th edn. McGraw-Hill (2004)

26. Liu, W., Zhu, S.S., Mundie, T., Krieger, U.: „Advanced block-chain architecture for e-Health systems", 2017 IEEE 19th IntConf e-Health networking. Appl. Serv. $\mathbf{1}$–$\mathbf{3}$, 2017 (2017)

27. Madhani, P.M.: Resource based view (RBV) of competitive advantage: an overview (March 26, 2010). In: Pankaj, M., (ed.) Resource Based View: Concepts and Practices, pp. 3–22. Icfai University Press, Hyderabad, India 2009 (2010). Available at SSRN: https://ssrn.com/abstract=1578704

28. Michael, S.C.: Entrepreneurial signaling to attract resources: the case of franchising. Manag. Decis. Econ. **30**, 405–422 (2009)
29. Park, J.-W., Yang, S.-B.: An empirical study on factors affecting blockchain start-ups' fundraising via initial coin offerings. In: Thirty Ninth International Conference on Information Systems. San Francisco (2018)
30. Risius, M., Spohrer, K.: A blockchain research framework. what we (don't) know, where we go from here, and how we will get there. Bus. Inf. Syst. Eng. **59**(6), 385–409 (2017)
31. Shin, L.: Here's The Man Who Created ICOs AndThis Is The New Token He's Backing (2017). https://www.forbes.com/sites/laurashin/2017/09/21/heres-the-man-who-created-icos-and-this-is-the-new-token-hes-backing/. Retrieved 11 Dec 2018
32. Spence, M.: Job market signaling. Quart. J. Econ. **87**, 355–374 (1973)
33. Swan, M.: Blockchain. Blueprint for a New Economy. O'Reilly Media (2015)
34. WHO.: WHA58.28 eHealth (2005). https://www.who.int/healthacademy/media/WHA58-28-en.pdf. Retrieved 13 Feb 2019

Chapter 13
Blockchain Track and Trace System (BTTS) for Pharmaceutical Supply Chain

Sine Canbolat, Özgür Ozan Şen, and Adnan Ozsoy

Abstract Blockchain technology supports highly sensitive data sharing with the improvement of security and privacy using cryptography. It is highly preferred in scenarios where the difficulty of tampering the data is critical. To manage supply chain in industrial distribution effectively, privacy, transparency, security, provenance and correctness of sensitive data need to be controlled between different participants. In particular, it has crucial importance in pharmacology that affects finance, investment and health of society. Thus in this study, we propose a blockchain track and trace system (BTTS) for improving management of pharmaceutical supply chain. We have offered state-of-the-art Hyperledger Fabric BTTS that distributed participants can control correctness of the medicine and protect highly sensitive data provenance for pharmaceutical supply chain. Proposed Hyperledger Fabric permissioned blockchain-based BTTS presents smart contract structure and protection for highly sensitive manufacturer's ownership data with less energy consuming than proof-of-work-based approaches.

13.1 Introduction

Blockchain technology has been introduced with the invention of best-known digital cryptocurrency system Bitcoin [1]. It is evaluated as the underlying technology behind cryptocurrencies, but it also gains an important place as a kind of distributed database that contains list of data. The data records in a distributed database are

S. Canbolat
University of Turkish Aeronautical Association, Ankara, Turkey
e-mail: sine.canbolat@ceng.thk.edu.com

Ö. O. Şen
Environmental Informatics in Helmholtz Center for Environmental Research, Leipzig, Germany
e-mail: oezguer-ozan.sen@ufz.de

A. Ozsoy (✉)
Hacettepe University, Ankara, Turkey
e-mail: adnan.ozsoy@hacettepe.edu.tr

© The Author(s), under exclusive license to Springer Nature Singapore Pte Ltd. 2021   227
S. Patnaik et al. (eds.), *Blockchain Technology and Innovations in Business Processes*,
Smart Innovation, Systems and Technologies 219,
https://doi.org/10.1007/978-981-33-6470-7_13

stored and referred as blocks that are linked together with cryptography. The key feature of blockchain technology is based on peer-to-peer, decentralized data management. While creating and reading operations are performed easily, altering and deleting records is highly difficult in blockchain [2]. These features are beneficial for immutable and critical data transactions.

Blockchain technology has a big potential to influence beyond the financial sector because it offers decentralized, secure ways to reach and share highly sensitive data. One such sector is pharmaceutical supply chain where there is a huge demand for new technologies for protection against counterfeit drugs. Counterfeit drugs not only dominate the financial gains of the producer but also effects human health. Radio-frequency Identification (RFID) tagging of product was offered for trusty drug tracking and tracing by Food and Drug Administration (FDA) [3]. Tracking of color, holograms, fingerprints, chemical markers, two-dimensional bar codes and labels in a drug were other technologies to authenticate safety and security of manufactured and distributed drugs. However, after medicine are expanded from the distributors, all this information is visible. Therefore, this medicine information could be cloned and medicine validation couldn't be guaranteed. Even if this information is cloned, drug safety needs to be satisfied. Thus pharmaceutical supply chain needs further improvements with new technologies. To improve pharmaceutical supply chain, traceability could directly increase public health and also reduce the financial loss in the medicine. Current safety and validation approaches like RFID need to be supported with the blockchain-based solutions that cover multiple distributed organizations.

In this study, we suggest Hyperledger Blockchain Track and Trace System (BTTS) for pharmaceutical supply chain. Following improvements are offered together in BTTS:

- The typical pharmaceutical supply chains compose the components, supplier, product manufacturer, distributer, and customers. Different permission levels are offered to control enrolled participants (government agency, manufacturer, hospital, pharmacy and patient) in the pharmaceutical supply chain. While both reading and writing operations are valid for government agency, manufacturer, hospital, and pharmacy, patients can only read from blockchain.
- Smart contracts are offered to improve safety and validation issues. Medicine transfer control over various types of smart contracts.
- Since manufacturers' endorsement is a highly sensitive data, medication ownership information is shared with any participants both government agency and related manufacturer permissions. This eliminates the financial concern of the manufacturer. Offered manufacturer endorsement protection is state-of-the-art approach in the blockchain-based pharmaceutical supply chain.

The rest of this chapter is presented as follows. Section 13.2 presents the previously published literature and reviews current solutions to healthcare area with blockchain-based solutions. In Sect. 13.3, we provide a background for Blockchain technologies. In Sect. 13.4 proposed BTTS is described in detail. Finally in Sect. 13.5 conclusion and future work can be found.

13.2   Related Works

Sensitive data production of the entire digital community is growing dramatically. Particularly, healthcare and biomedical ecosystem have a big potential to store and share distributed sensitive data. Additionally, making use of this data is important blockchain features like security and efficiency can provide a smooth adaption for solutions that were not possible before. A few research and review articles in regard to adopt blockchain features to healthcare and biomedical ecosystem indicate potentials and some solutions for access control and medical researches.

Kuo et al. [4] focused on some challenges of implementing blockchain solutions to biomedical/healthcare applications with the possible improvement ways. These issues were listed as highly sensitive personal data, realtime speed, scalability and existence of 51% attack. Encrypting highly sensitive data on the network offered to deal with transparency issue of highly sensitive health data. This research stated that blockchain technology had great potential to offer novel biomedical and healthcare applications.

According to Linn et al. [5] study, storing all health data, tracking in real-time and supporting secure access permission could be beneficial in terms of fast and transparent development of blockchain features implementation of healthcare system. Blockchain implementation covered all the patients' health data. In this study, scalability issue was met with implementation of a data repository that was called as data lake. Moreover, only data owner gave permission to access personal data by access control policies. Privacy and anonymity were main concern when sharing healthcare-sensitive data.

In the study of Peterson et al. [6], three assumptions exist for sharing data on blockchain network. Data could be shared and received if and only if auditability and security were assured. Another crucial issue was based on access control to patient data. Patient authorizes participants to reach their data through smart contract implementation.

Yue et al. [7] proposed Healthcare Data Gateway (HDG) application in order to reduce privacy risk with the implementation of three layers that were called as storage layer, data management layer and data usage layer. The aim of this application was based on offering decentralized platform and smart healthcare architecture. Proposed architecture offered secure ways to share personnel health data.

Azaria et al. [8] implemented MedRec system to support effective sharing policy between multisources. Medical record automation and tracking were supported via Ethereum blockchain smart contracts. DNS-like implementation was preferred for the identity confirmation. To share various metadata fields within a single record, additional restrictions could be adopted with smart contract.

Dagher et al. adopted [9] blockchain features to share healthcare data in a decentralized approach that is called as Ancile. Ancile framework covered six smart contracts for user registration, certain overwrite, maintain blockchain mining, classify different nodes' level, remove risk duplicated registration in the system, maintain history of relationship and holding address of proxy nodes. Even if main domain of presented works was based on healthcare operations and personal health records, these solutions could also bring light to supply chain and pharmaceutical researches. Because these researches present possible solutions for controlling highly sensitive data that is also main concern of manufacturer's endorsement in pharmaceutical supply chain.

Blockchain features can be implemented for pharmaceutical solutions. To fight with counterfeit drug problem, there exist some state-of-the-art solutions. Mettler [10] applied blockchain features to mark counterfeit drugs. Researches pointed out benefits of blockchain adaptation impact not only health data management and researches but also for the pharmaceutical industry. It was indicated that Bitcoin transaction technology could be also adopted for monitoring production process of drugs. While potentials of blockchain-based solutions to reduce counterfeit drugs explained theoretically, implementation results and technical explanations were needed for further improvements.

Schöner et al. [11] presented LifeCrypter as a prototype solution to improve security for drug supply chain. Integrity, traceability and transparency were verified with the help of smart contract implementation to drug chain process globally. In the LifeCrypter, it was indicated that only trusted organizations could write to blockchain. Manufacturers and patient could reach the history. They attached a unique tag for each medical item and verified a trusted chain by smart contracts implementation for all stages. However, the smart contract did not presented for further studies and implementations.

Another potential of applying blockchain technology to drug supply chain was to increase information sharing between investors and research companies. Clauson et al. [12] suggested that blockchain features could be beneficial to improve data security, integrity, provenience and functionality of health supply chain. Product identification, tracing, verification, detection and response, recall management, notification and information requirements could be met with blockchain solutions [12].

Toyoda et al. [13] proposed blockchain features implementation for product ownership management system to reduce counterfeits drugs in the post supply chain. According to this study, to avoid cloned Radio Frequency Identification (RFID) tags, a blockchain-based decentralized application Ethereum platform was offered in post supply chain process. The study assumed that reselling of products was allowed in the post supply chain. An Electronic Product Code (EPC) was embedded for all products, and these products are delivered through supply chain organizations. In this supply chain process, each organization control RFID tags and added required data into tags. This process helped the further organizations to verify the product. In any unexpected situation, this product evaluated as counterfeits. Proof of possession products was implemented for the post supply chain process. First ownership was belong to valid manufacturers. With the help of developed system customers could identify counterfeit drugs. According to performance evaluation, the proposed system needed less than US$1 to manage products with nearly six transfers. This study referred some open issues such as security issue, transfer reward quantity and protect manufacturer's endorsement.

Tseng et al. [14] applied Gcoin blockchain's double-spending prevention mechanism which is Consortium Proof-of-Work approach to solve counterfeit-drug problem. Each seller and buyer had own public key in the drug supply chain like in Bitcoin system. Using batch or serial number, all drug information was offered to generate hash number as the public key. With the help of this public key, a Quick Response (QR) code was generated to identify drugs. While manufacturers and government agencies were responsible for verifying transactions and generating blocks, wholesalers, hospitals or third parties are in charge for storing a backup of historical transactions. Furthermore, pharmacies and consumers were evaluated as the authority to implement transactions. Multisignature design was offered to support identical digital sign for delivering, transacting, and inspecting of the drugs in the chain. According to this study, average speed is nearly 17.5–26 transactions per second. This research also has an open issue about manufacturer's endorsement protection. When focuses on Bitcoin-based GCoin and Ethereum blockchain technologies, there exists next-generation blockchain technology with less energy consuming, high transaction and scalability.

The existing studies described potential of blockchain-based solutions in health care and pharmacy. These current solutions have some limitations in terms of describing/protecting highly sensitive data in pharmaceutical supply chain and energy consumption and/or transaction throughput in blockchain. To meet these current requirements, in this study sensitive data protection approach is offered for pharmaceutical supply chain and also Hyperledger Fabric framework is preferred to support high throughput and less energy consumption.

13.3 Overview of Bitcoin, Blockchain Platforms and Smart Contracts

The full description of how Blockchain works and the details of the Bitcoin structure are described in detail in many papers and venues; thus, we do not cover the basic introduction in this study.

However, to be able to choose the best option for pharmaceutical supply chain, we need to look more deeply into how transactions handled in different frameworks. In Bitcoin, transactions pass an asset from one owner to another, so the asset never moves, but only owner changes for a certain asset. Bitcoin refers these assets Unspent Transaction Outputs (UTXO) which refers all assets that is not spent yet. UTXO model holds a public key and also stack-based programming language. Additionally, UTXO has two conditions that is named as spent and unspent. Hence, multistage, stateful contracts couldn't be built by Bitcoin scripting language. Furthermore, the building components of a block which are nonce, the timestamp and previous block hash are not accessible by UTXO model. This brings up some limitations to applications. Additionally, avoiding from loops in Bitcoin-based scripting language is also a limitation.

To overcome these limitations, the Ethereum proposed smart contract model which allows anyone to create transaction rules and ownership of any asset. Ethereum blockchain expands on Bitcoin that is decentralized platform with smart contract implementation [15]. Source code of a smart contract is invoked by user's transaction. This contract can return a value to the sender, or invoke another routine and may lead to another transaction. Receiver of an Ethereum message, if it exists in a contract, has an option to return an answer. In order to preserve from infinite loops in Ethereum script, there exist a price to run the code, named gas price, which is a fee for miners to pay each computational step. Thus, these steps consume gas for a transaction's executions [16].

Bitcoin and Ethereum are public blockchains which means all the transactions are public and anyone can join. However, not every scenario is suitable for such openness and publicity. To address this problem, the concept of private blockchains arises. In a private blockchain, the identities of the nodes are usually known by other peers. The entrance to the network is limited. Hyperledger, which is built as an umbrella project of open source private blockchain by the Linux Foundation, IBM and Intel, is one of the most preferred private blockchain platform. These blockchain platforms have strengths and weaknesses in theory and application. Ethereum has also same consensus mechanism as Bitcoin. In Bitcoin-based blockchain, energy consumption could reach high levels [17].

Next-generation blockchain technology that is called Hyperledger Fabric performs better than Ethereum in terms of energy consumption. Hyperledger Fabric offers private network and modular architecture with consensus protocols. It also supports smart contract implementation as well.

13.4 The Proposal Model

13.4.1 Structure of Blockchain Track and Trace System (BTTS)

To increase transparency and eliminate counterfeits drugs in the supply chain, we propose to use Hyperledger Fabric platform as an efficient and effective solution. The offered BTTS can track and verify with the help of blockchain technology. Each transfer is controlled by own addresses of medicine, sender and receiver in the pharmaceutical supply chain. To support track and trace system, all required medicine data (medicine owner, serial number, name, amount), RFID information (temperature, humidity), transfer state, transfer number, expected location, current location, sender/receiver approvement states, current pair approvement states are recorded in our proposed BTTS model. QR code is used for medicine validation.

In pharmaceutical supply chain, different types of participants exist that named as government agencies, manufacturers, distributors, hospitals/pharmacies and patients (end users). Pharmaceutical supply chain requires the approval and control of an authorized participant. Therefore, the government agencies as an authorized participant assign manufacturers, distributors, hospitals/pharmacies and patients in the permissioned BTTS. Sender (manufacturers, distributors, or hospitals/pharmacies) checks location of medicine. Receiver (distributors, hospitals/pharmacies or patients) controls both medicine information, transfer state, RFID information, expected location, current location and transfer number to validate medicine. Manufacturers create a hash value that covers medicine name, medicine serial number, medicine private key, medicine quantity and all required information of medicine with manufacturer's private keys. Hashing with private key and 2 of 2 multi-signature design are offered to protect inference of confidential business information of companies like endorsement. Only permissioned participants can reach manufacturer ownership information by authorization of both government agency and manufacturer. Besides, senders and receivers create hash value with their private key respectively for protecting sensitive information. 1 of 2 multi-signature design is suggested to reach data of sender or receiver. Sender/receiver data access authorization is given by government agency or sender/receiver permission. In Fig. 13.1, structure of Blockchain Track and Trace System (BTTS) for pharmaceutical supply chain is described according to involved participants. Creating hash values with private keys and multi-signature design to protect highly sensitive data are two main concerns in BTTS.

Medicine transaction is started when manufacturer transfers their medicine to the receiver and transfer state is turned to active from passive. Medicine public key and encrypted medicine data are available for all participants.

Receiver checks manufacturer ownership with the help of medicine public key and medicine hash value. The compatible hash value decrypted with 2 of 2 multi-signature approvement of government agency and the valid manufacturer. All manufacturers attempt to decrypt the broadcasted hash value with their company private key along with government agency. The manufacturer which can decrypt the hash will be able

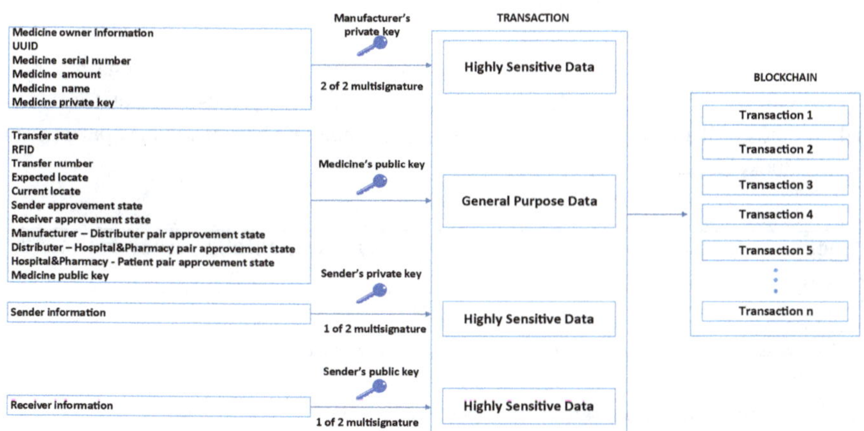

Fig. 13.1 Structure of Blockchain Track and Trace System (BTTS)

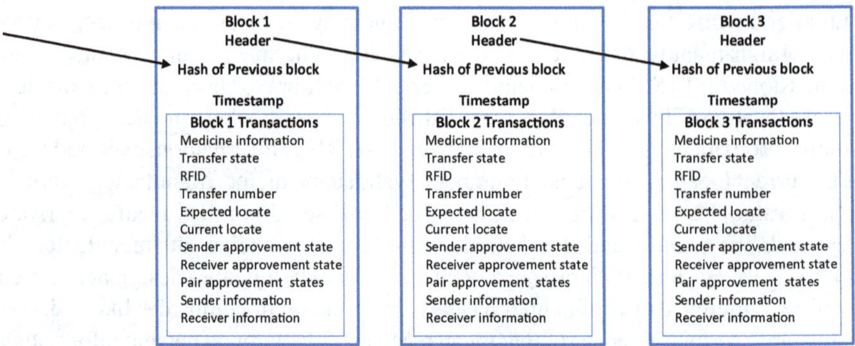

Fig. 13.2 Flow of the offered BTTS for pharmaceutical supply chain

to access and retrieve the medicine private key. If the manufacturer finds out the medicine private key ownership belongs to itself, then it may return a valid result through the blockchain. In this way, manufacturer ownership is confirmed. After the ownership validation process, receiver also verifies RFID information, expected location, current location and transfer number to validate medicine. If medicine passed these controls then transfer state, transfer number, expected location, current location, sender/receiver approvement states, current pair approvement and sender/receiver information are updated. Otherwise medicine is assigned as counterfeit. At the last stage, patient can verify medicine public key's QR code. With the help of the QR code, patient can check authorized approvals. In Fig. 13.2, flow of the offered BTTS for pharmaceutical supply chain is presented.

13.4.2   The Proposed Smart Contracts

Smart contract is presented in order to control pharmaceutical supply chain. Algorithm 1 shows the pseudo-code of createParticipant() that creates manufacturer, distributor, hospital and patient when government agencies decide to add new participant to the blockchain.

Algorithm 1 Pseudo-code of *createParticipant()* which enrolls participant into the Blockchain

---

1: Inputs: *UUID, info, type*
2: Output: *a boolean*
3: **if** the *message sender*'s participant type is NOT *"government agency"* **then**
4:     return *false*
5: **else**
6:     enroll participant to the blockchain into the list *participant[UUID]*
7:     return *false*

Algorithm 2 describes the createMedicine() which is for generating medicine with ownerName, serialNumber, UUID and all required medicine information.

Algorithm 2 Pseudo-code of *createMedicine()* which enrolls medicine into the Blockchain

---

1: Inputs: *ownerName, serialNumber, UUID, info, manufacturerInfo*
2: Output: *a boolean*
3: **if** the *medicineStore* list OWNS the product with *UUID* **then**
4:    return *false*
5: **if** the *serialNumber* does NOT belong to *manufacturerInfo.address* **then**
6:    return *false*
7: set *status* to *"obtained"*
8: set *owner* to *message sender*
9: set *currentLocation* to *message sender*
10: set *transferCount* to *"0"*
11: enroll medicine to blockchain into the list *medicineStore[UUID]*
12: return *true*

Algorithm 3 describes transferMedicine() which is for starting the medicine transaction between sender and receiver. The function gets address of recipient and medicineUUID and controls correctness of receiver from the blockchain. If it is the expected receiver, transfer state is updated. Otherwise transaction cannot be started by any participant. Patient cannot transfer medicine based on offered BTTS model.

Algorithm 3 Pseudo-code of *transferMedicine()* which starts a transfer from medicine's current location to the expected location

---

1: Inputs: *recipientAddress, medicineUUID*
2: Output: *a boolean*
3: **if** the product with *UUID* does NOT exist **then**
4:    return *false*
5: **if** the medicine with *UUID* NOT owned by *message sender* **then**
6:    return *false*
7: **if** the *message sender* is a *"patient"* **then**
8:    return *false*
9: set *status* to *"transferred"*
10: set *expected* to *recipientAddress*
11: update *approveList* with *message sender address*
12: update *senderInfo*
13: return *true*

Algorithm 4 describes acceptMedicine() which is for confirming the arrival medicine with the help of sender, receiver and manufacturer cooperation. Medicine correctness is controlled by both manufacturer and government agency authentication.

Algorithm 4 Pseudo-code of *acceptMedicine()* which concludes the pending transfer from medicine's previous location

---

1: Inputs: *shipperAddress, medicineUUID*
2: Output: *a boolean*
3: **if** *approveList* of the medicine contains any unapproved field **then**
4:     return *false*
5: **if** the *currentLocation* of the medicine is NOT *shipperAddress* **then**
6:     return *false*
7: **if** the *expectedLocation* of the medicine is NOT *recipientAddress* **then**
8:     return *false*
9: **if** the *status* of the medicine is NOT *"transferring"* **then**
10:     return *false*
11: **if** the *medicineInfo.temp* or the *medicineInfo.humidity* of the medicine is NOT appropriate **then**
12:     return *false*
13: **if** *approveList.length* > *"TRANSACTION_NUMBER"* **then**
14:     return *false*
15: set *status* to *"obtained"*
16: set *currentLocation* to *recipientAddress*
17: set *expectedLocation* with *null*
18: update *approveList* with *recipientAddress*
19: update *recipientInfo*
20: return *true*

13.5   Conclusion and Future Works

We introduced a novel blockchain-based solution to improve track and trace system for pharmaceutical supply chain. Blockchain distributed ledger technology can enhance the pharmaceutical supply chain management to improve public health and avoid financial lost. Firstly, we identified benefits of blockchain technology compared to current solutions of pharmaceutical supply chain management effectively. Then proposed Blockchain Track and Trace System (BTTS) model was described to support manufacturers' privacy and also medicine traceability. The offered system needs to be implemented for further studies. Besides, the offered system could be covered for secondary parties such as the Internet markets. Hyperledger Fabric blockchain-based state-of-the-art solution, BTTS prototype, has a big potential to reduce energy consuming, increase transaction throughput and reduce counterfeit drugs. The future of blockchain technology in pharmaceutical supply is still developing because technology is early stages in the presented domain.

Funding Statement This research received no specific grant from any funding agency in the public, commercial or not-for-profit sectors.

Competing Interests Statement Authors have no competing interests in this research.

Contributorship Statement All authors conceived of the presented idea, developed the theory and performed the computations. All authors discussed the results and contributed to the final manuscript.

References

1. Nakamoto, S.: Bitcoin: A peer-to-peer electronic cash system (2008)
2. Bigchaindb 2.0 the blockchain database. Whitepaper (2018). https://www.bigchaindb.com/whitepaper/bigchaindb-whitepaper.pdf
3. Combating counterfeit drugs: A report of the food and drug administration (2019). https://www.fda.gov/downloads/Drugs/DrugSafety/UCM169880.pdf
4. Kuo, T.T., Kim, H.E., Ohno-Machado, L.: Blockchain distributed ledger technologies for biomedical and health care applications. J. Am. Med. Inform. Assoc. **24**(6), 1211–1220 (2017)
5. Linn, L.A., Koo, M.B.: Blockchain for health data and its potential use in health it and health care related research. In: ONC/NIST Use of Blockchain for Healthcare and Research Workshop. Gaithersburg, Maryland, United States: ONC/NIST (2016)
6. Peterson, K., Deeduvanu, R., Kanjamala, P., Boles, K.: A blockchain-based approach to health information exchange networks. In: Proceedings of NIST Workshop Blockchain Healthcare, vol. 1, pp. 1–10 (2016)
7. Yue, X., Wang, H., Jin, D., Li, M., Jiang, W.: Healthcare data gateways: found healthcare intelligence on blockchain with novel privacy risk control. J. Med. Syst. **40**(10), 218 (2016)
8. Azaria, A., Ekblaw, A., Vieira, T., Lippman, A.: Medrec: Using blockchain for medical data access and permission management. In: 2016 2nd International Conference on Open and Big Data (OBD), pp. 25–30. IEEE, New York (2016)
9. Dagher, G.G., Mohler, J., Milojkovic, M., Marella, P.B.: Ancile: Privacy-preserving framework for access control and interoperability of electronic health records using blockchain technology. Sustain. Cities Soc. **39**, 283–297 (2018)
10. Mettler, M.: Blockchain technology in healthcare: The revolution starts here. In: 2016 IEEE 18th International Conference on e-Health Networking, Applications and Services (Healthcom), pp. 1–3. IEEE, New York (2016)
11. Schöner, M.M., Kourouklis, D., Sandner, P., Gonzalez, E., Förster, J.: Blockchain Technology in the Pharmaceutical Industry. Frankfurt School Blockchain Center, Frankfurt, Germany (2017)
12. Clauson, K.A., Breeden, E.A., Davidson, C., Mackey, T.K.: Leveraging blockchain technology to enhance supply chain management in healthcare. Blockchain in Healthcare Today (2018)
13. Toyoda, K., Mathiopoulos, P.T., Sasase, I., Ohtsuki, T.: A novel blockchain-based product ownership management system (POMS) for anti-counterfeits in the post supply chain. IEEE Access **5**, 17465–17477 (2017)
14. Tseng, J.H., Liao, Y.C., Chong, B., Liao, S.W.: Governance on the drug supply chain via gcoin blockchain. Int. J. Environ. Res. Public Health **15**(6), 1055 (2018)
15. Wood, G.: Ethereum: A secure decentralised generalised transaction ledger. Ethereum Project Yellow Paper **151**, 1–32 (2014)
16. Delmolino, K., Arnett, M., Kosba, A., Miller, A., Shi, E.: Step by step towards creating a safe smart contract: Lessons and insights from a cryptocurrency lab. In: International Conference on Financial Cryptography and Data Security, pp. 79–94. Springer, Berlin (2016)
17. Symitsi, E., Chalvatzis, K.J.: Return, volatility and shock spillovers of bitcoin with energy and technology companies. Econ. Lett. **170**, 127–130 (2018)

CPSIA information can be obtained
at www.ICGtesting.com
Printed in the USA
LVHW080414060421
683527LV00002B/110